International Conference on Gravitation Cosmology and Continuum Mechanics 2016

Dedicated to the Centenary of Kirill Petrovich Staniukovich

Journal of Physics: Conference Series Volume 731

Moscow, Russia
3 - 4 March 2016

ISBN: 978-1-5108-2729-5
ISSN: 1742-6588

Printed from e-media with permission by:

Curran Associates, Inc.
57 Morehouse Lane
Red Hook, NY 12571

proceedings.com

Some format issues inherent in the e-media version may also appear in this print version.

Printed by Curran Associates, Inc. (2016)

For permission requests, please contact the Institute of Physics
at the address below.

Institute of Physics
Dirac House, Temple Back
Bristol BS1 6BE UK

Phone: 44 1 17 929 7481
Fax: 44 1 17 920 0979

techtracking@iop.org

Additional copies of this publication are available from:

Curran Associates, Inc.
57 Morehouse Lane
Red Hook, NY 12571 USA
Phone: 845-758-0400
Fax: 845-758-2633
Email: curran@proceedings.com
Web: www.proceedings.com

Table of contents

Volume 731

International Conference on Gravitation Cosmology and Continuum Mechanics (dedicated to the centenary of Kirill Petrovich Staniukovich)
3–4 March 2016, Moscow, Russia

Accepted papers received: 8 July 2016
Published online: 26 July 2016

Preface

Papers

INTRODUCTION

International Conference – Gravitation, Cosmology and Mechanics of Continuous Environments (devoted to the 100th anniversary K.P. Stanyukovich's birth) was held in Bauman Moscow State Technical University on the 3th and 4th of March. More than 100 papers were presented by K.P. Stanyukovich's students, faculty, various universities staff and representatives of the Russian Academy of Sciences.

Kirill Petrovich Stanyukovich (3 March, 1916 – 4 June, 1989) – an outstanding physicist, mathematician and engineer made a significant contribution to the development of various fields of science: gas dynamics, physics of explosion, magnetic hydrodynamics, astronomy. He developed a hydrodynamic model of gravity, the theory of gravity with a variation of the effective gravitational constant, with a variable number of particles, he offered one of the first Universe evolution scenarios from the initial vacuum stage.

Kirill Petrovich was the author of a number of inventions and of 330 scientific and popular-science works. He came from an old noble family – the Stanyukovich. (The famous writer and marine painter Konstantin Stanyukovich was his granduncle).

Not yet having finished school, K.P. Stanyukovich got seriously interested in astronomy and became a member of the observers' team of the Moscow Society of Astronomy Fanciers (MSAF) where he met Leonid Alekseevich Kulik, the famous Tunguska meteorite investigator and one of the national meteoritics founders. In 1932 the first K.P. Stanyukovich's research article on meteor astronomy (in co-authorship with I.E. Vasilyev) "Lyrids in 1930" was published in the "Bulletin of MSAF Observers. In 1931 K.P. Stanyukovich took part in the MSAF meteors observation expedition on the Karadag scientific station in the Crimea, and in 1932 proposed a new method for estimating the meteor extinction height, based on an empirical relationship between meteor brightness and path length. In the same year he was involved in the processing of bright bolide observations over the town Shepetovka (Ukraine), and, in co-authorship with V.V. Fedynskiy, he carried out experiments on photographing meteors with obturator camera. Successful basic images of bright meteor in August 14, 1932 allowed them to calculate not only the height, path length and meteor speed but also the upper atmosphere basic parameters in the span place.

In 1933 he proposed one more method of meteor astronomy - the comparison method - according to which one of the meteor stream parameters made it possible to determine the parameters of another meteor. In 1934, a group of MSAF observers led by K.P. Stanyukovich received the first in the USSR and the 35th in the world bright meteor spectrum with 47 spectral lines. And in 1936, 20-year-old K.P. Stanyukovich was charged to prepare and conduct the solar eclipse observations on board of an airplane that helped to minimize interference from the atmosphere and increase the total phase observations time. The observation results were published in the "Mirovedenie" journal in the same year.

In 1936-1939 K.P. Stanyukovich worked as a researcher of the Committee on Meteorites of the Academy of Sciences of the USSR, and in the same time studied at MSU Faculty of Mechanics and Mathematics correspondence department (specialty - "Astronomy"). In 1939 he was drafted into the Red Army, where he served in Poltava aviation courses, in the rank of junior commander, but in May 1941 he was discharged because of an eye disease, which enabled him to pass state exams and get a university diploma.

The Great Patriotic War brought significant changes in K.P. Stanyukovich scientific specialization and he was actively engaged in the gas dynamics and physics of explosion problems. There is no reliable information what the father was doing in the first months of the war, and family legends about his short service in the bombers haven't had documentary evidence yet, but it is known that in September 1941 he was working on specialized subjects following the instructions of the State Defense Committee commissioner C.V. Kaftanov, first in People's Commissariat of Mortar Weapons, and then - in the Engineering Committee of the Red Army. We also know that in 1945 with the rank of engineer–major in a group of military experts he was sent to Tallinn and Berlin to study German explosives and ammunition. The published biography of K.P. Stanyukovich is silent about these trips but his son Andrew K. Stanyukovich remembers well his father's stories about Russian soldiers' autographs on the Reichstag ruins and about the group return from Berlin to Moscow on "Douglas", where during the flight the passengers and the crew, fighting cold, warmed themselves with alcohol they had, with the result that instead of Moscow they landed at Thule.

In 1944 K.P. Stanyukovich defended his candidate's thesis, and in 1946 - a doctoral thesis. After the war he worked as a senior teacher at the F.E. Dzershinskiy Artillery Academy, then he worked as a head of the laboratory at the Institute of Problems of Chemical Physics of the Academy of Sciences of the USSR and then as a head of the department in the Bauman Moscow Higher Technical School, apparently without interrupting his cooperation with the so-called "Kaftanov Commission ", resulting in his theoretical solution (co-authored with Landau) on the problem of a convergent shock wave (implosion), which was the basis of one of initiation embodiments of a nuclear explosion in the Soviet atomic bomb development, first tested in 1949.

In 1950 - 1960 K.P. Stanyukovich, continued theoretical developments in the field of gas dynamics and magneto hydrodynamics, he related them to the problems of meteor astronomy and meteoritics, such as the theory of cratering in a collision of meteoroids with planet surface, calculation of planetary substance fragment emission at cratering, theory of thermal explosion at the ice body hypersonic flight in the atmosphere (with reference to the comet's version of the Tunguska phenomenon), description of air shock waves system during the flight of large bolides and others. Independently and in co-authorship with V.V. Fedynskiy, F.A. Baum, G.I. Pokrovskiy, S.K. Vsekhsvyatskiy, V.A. Bronshten, A. K.Mukhamedzhanov, V.P. Shalimov and others) he worked on some problems of cosmogony, using higher mathematics and gas dynamics methods. In 1952 he was awarded the title of professor. In 1958, in co-authorship with F.A. Baum and S.A. Kaplan he published the monograph "Introduction to the cosmic gas dynamics".

Among K.P. Stanyukovich's fundamental works in the field of physical gas dynamics and physics of explosion we should mention the monograph "Unsteady motion of a continuous medium (1955, 1960, 1971) and" Physics of Explosion "(1959, 1975, 2002, co-authored), that withstood several editions. His active lecture and popularization activity in the field of astronomy, cosmology, rocket and space technology are widely known. Back in the late 1950s K.P. Stanyukovich became interested in the universe structure and evolution problems and this came to reality in the monograph "The gravitational field and elementary particles" (1965).

Since 1967 until his death K.P. Stanyukovich scientific activity took place mainly in the system of the State Standard of the USSR, where he formed a unique team of theoretical physicists specializing in astrophysics, cosmology and gravitation fields.

In the years 1969-1972 K.P. Stanyukovich was a chairman of the Moscow branch of the All-Union Astronomical and Geodetic Society. In 1974 he was awarded the honorary title "Honored Worker of Science and Technology of the RSFSR" and in 1981 for his work on special subjects he was awarded the State Prize of the USSR.

In 1983 he co-authored with V.N. Melnikov published a monograph "Hydrodynamics, Fields and Constants in the Theory of Gravitation".

This publication contains selected papers presented at the conference.

Participants

Presidium

Participants

Journal of Physics: Conference Series **731** (2016) 011002 doi:10.1088/1742-6596/731/1/011002

Organizing Committee of the Conference

A.A. Aleksandrov (Chairman) – Dr.Sci. (Engineering), Professor, Rector of Bauman MSTU

V.N. Zimin – Dr.Sci. (Engineering), First Pro-Rector (Research), Bauman MSTU

V.O. Gladyshev – Dr. Sci. (Phys., Math.), Head of NUK FN, Bauman MSTU

V.T. Kalugin – Dr. Sci. (Engineering), Head of NUK SM, Bauman MSTU

A.N. Morozov – Dr. Sci. (Phys., Math.), Professor, Head of Dept. FN-4, Bauman MSTU

Ju.I. Dimitrienko – Dr. Sci. (Phys., Math.), Professor, Head of Dept. FN-11, Bauman MSTU

L.P. Orlenko – Dr. Sci. (Engineering), Professor, Dept. SM-4, Bauman MSTU

V.V. Zelencov – PhD (Engineering), lecturer, Dept. SM-6, Bauman MSTU

V.N. Mel'nikov – Dr. Sci. (Phys., Math.), Professor, Head of the Centre of Gravitation and Fundamental Metrology, VNIIMS; Deputy director of the Institute of Gravitation and Cosmology, PFUR; President of RGS

K.A. Bronnikov – Dr. Sci. (Phys., Math.), Chief researcher, Centre of Gravitation and Fundamental Metrology, VNIIMS; Professor Institute of Gravitation and Cosmology, PFUR

N.N. Sysoev – Dr. Sci. (Phys., Math.), Professor, Dean of the Faculty of Physics of Lomonosov Moscow State University

Ju.V. Gerasimov – PhD (Engineering), lecturer, Dept. Of Physics, Bauman MSTU

P.N. Antonyuk – PhD (Engineering), Dept. FN-11, Bauman MSTU – Scientific Secretary of the conference

A.P. Fournier-Sikr – Dr., European Space Agency, France

G.M. Webb – Prof., Commercial Space Technologies Ltd., London, United Kingdom

Academic Committee of the conference

A.I. Leont'ev (Chairman) – Academician of RAS, Bauman MSTU

V.O. Gladyshev – Dr. Sci. (Phys., Math.), Head of NUK FN Bauman MSTU

V.T. Kalugin – Dr.Sci. (Engineering), Head of NUK SM Bauman MSTU

V.V. Selivanov – Dr.Sci. (Engineering), Professor, Head of Dept. SM-4, Bauman MSTU

M.I. Kiselev – Dr. Sci. (Phys., Math.), Professor, Dept. MT-4, Bauman MSTU

R.Z. Kavtaradze – Dr. Sci. (Engineering), Professor, Bauman MSTU

V.N. Mel'nikov – Dr. Sci. (Phys., Math.), Professor, Head of the Centre of Gravitation and Fundamental Metrology, VNIIMS; Deputy director of the Institute of Gravitation and Cosmology, PFUR; President of RGS

Ju.S. Vladimirov – Dr. Sci. (Phys., Math.), Professor, Lomonosov MSU

D.V. Gal'cov – Dr. Sci. (Phys., Math.), Professor, Lomonosov MSU

A.P. Efremov – Dr. Sci. (Phys., Math.), Professor, First Pro-Rector (Education), PFUR, Director of the Institute of Gravitation and Cosmology of PFUR

Ju.P. Rybakov – Dr. Sci. (Phys., Math.), Professor, Head of Dept. of Theor. Physics and Mechanics, PFUR

Ju.G. Rudoj – Dr. Sci. (Phys., Math.), Professor, Dept. of Theor. Physics and Mechanics, PFUR

V.P. Vizgin – Dr. Sci. (Phys., Math.), Head of sector, S. Vavilov IHST , RAS

V.L. Gvozdeckij – PhD, Head of Dept., Head of sector, S. Vavilov IHST , RAS

G.Yu. Bogoslovsky – Dr. Sci. (Phys., Math.), Professor, Skobeltsyn Institute of Nuclear Physics, Lomonosov Moscow State University

P. Fournier-Sikr – Dr., European Space Agency, France

G.M. Webb – Prof., Commercial Space Technologies Ltd., London, United Kingdom

P. Rowlands – Prof. University of Liverpool, Liverpool, United Kingdom

M. Abishev – Prof., Al-Farabi Kazakh National University, Head of Theoretical and Nuclear Physics Department, Almaty, Kazakhstan

E. Chubaryan – Prof., Yerevan State University, Head of Laboratory, Yerevan, Republic of Armenia

A.V. Minkevich – Prof., Belarusian State University, Minsk, Belarus and Warmia and Mazury University in Olsztyn, Poland

A.I. Zhuk – Prof., Odessa National University, Odessa, Ukraine

Journal of Physics: Conference Series **731** (2016) 011004 doi:10.1088/1742-6596/731/1/011004

Peer review statement

All papers published in this volume of *Journal of Physics: Conference Series* have been peer reviewed through processes administered by the proceedings Editors. Reviews were conducted by expert referees to the professional and scientific standards expected of a proceedings journal published by IOP Publishing.

Journal of Physics: Conference Series **731** (2016) 012001 doi:10.1088/1742-6596/731/1/012001

Mathematics of oscillations and waves

P N Antonyuk[1,2]
[1] Docent, Bauman Moscow State Technical University, Moscow, Russia
[2] Senior Researcher, S.I. Vavilov Institute of History of Science and Technology of the Russian Academy of Sciences (IHST RAS), Moscow, Russia
E-mail: pavera@bk.ru

Abstract. Analytical definition of the sine function and the number π which is related to sine function allows to understand, how harmonic oscillations and waves appear. The functional equation for the sine is received. The new fast algorithm of calculation of the π number is formulated. In the elementary case oscillations and waves are harmonious or sinusoidal. The sine function appears not accidentally. This function can be defined as the solution of the functional equation characterizing periodic properties of oscillations and waves.

1. The functional equation

We will consider the functional equation

$$f(x+p)=-f(x)\ ,\ p>0,\ x\in \mathbf{R}\ ,$$

in which p is half-period of the f function. The half-period p unambiguously defines period $2p$ of a periodic function.

The general solution of the equation in the class of analytical functions bounded on the real straight line is a three-parameter family of sinusoids

$$\begin{cases} f(x)=a\sin(bx+c); \\ p=\dfrac{\pi}{b}, \end{cases}$$

where a,b,c are free parameters.

The particular solution of the equation

$$f(x)=r\sin\left[\frac{\pi}{p}(x-q)\right]$$

follows from the system

$$\begin{cases} f(x+p)=-f(x); \\ f(q)=0; \\ f\left(q+\dfrac{p}{2}\right)=r \end{cases}$$

with given parameters p,q,r.

Finally, the system

Journal of Physics: Conference Series **731** (2016) 012001 doi:10.1088/1742-6596/731/1/012001

$$\begin{cases} f(x+p) = -f(x); \\ f(0) = 0; \\ f'(0) = 1; \\ f\left(\frac{p}{2}\right) = 1 \end{cases}$$

gives the only solution

$$\begin{cases} f(x) = sin\ x, \\ p = \pi. \end{cases}$$

Thereby, the functional equation allows to define the sine in a new way.

2. The number π

Important statements for the number π also follow from the functional equation.
We will consider the mapping

$$F = \{\sin,\ \arcsin\} = \sin\circ\arcsin + \arcsin\circ\sin = \mathrm{id} + \arcsin\circ\sin$$

defined as anti-commutator of sine and arcsine. By means of the symbol id here is designated identical function.

Defenition. The number π is a stable fixed point of mapping F of a unit disc $|z-3| \leq 1$ of complex plane onto itself.

Contracted mapping F has the only one fixed point in the disc. Stability of the point means that $F(\pi+\varepsilon) = \pi$ for small absolute values of arbitrary complex number ε.

Theorem. $\pi \equiv F(z)$, where $|z-3| \leq 1$.

Corollary. $\pi = F(3)$ or $\pi = 3 + \arcsin\ \sin 3$.

The number π which is the $z = F(z)$ equation root we can find by the successive approximations method. The necessary condition of the method's convergence is defined by a double inequality $-1 < F(z) < 1$. As $F'(z) \equiv 0$ in a unit disc, so the method is reduced to one iteration $z \mapsto \pi$ that means infinitely high rate of convergence of successive approximations to a fixed point. In other words, mapping F is greatly contracted. Therefore, the theorem gives new fast algorithm for the number π calculation.

3. Rate of convergence

Rate of convergence of iterative process shows how quickly the sequence of approximations converge to the limit: the number of the correct decimal signs of one iteration approach increases in k times. For comparison, we will give values of convergence rate for the most important iterative methods of the solution of the algebraic and transcendental equations:
$k \geq 1$ Successive approximations method,
$k = \tau$ Secant method or chord method ($\tau = 1{,}618...$ – a golden section),
$k = 2$ Newton's method or method of tangents,
$k = 3$ Halley's method,
$k = \infty$ Chebyshev's method,
$k = \infty$ New fast algorithm of the number π calculation.

One iteration gives exact value of a limit when there is the infinitely large rate of convergence.

4. Calculations of the number

π by means of formulae

$$\pi = \{sin,\ arcsin\}(2) = \{sin,\ arcsin\}(3) = \{sin,\ arcsin\}(4) =$$
$$= \{sin,\ arcsin\}(3+i) = \{sin,\ arcsin\}(\tau) = \{sin,\ arcsin\}(e),$$

subject to preservation N decimal digits for all received intermediate numbers, give N correct signs of the number π for each formula. Here $e = 2,718...$ is the natural logarithmic base, i is the imaginary unit. It speaks about the stability of computing process which is coming from stability of the fixed point of mapping F.

5. The law of similarity

Let the number π be calculated with a precision N decimal signs by means of our algorithm, let t be time of calculation of the number π, t_* be characteristic time of calculation. Computational experiments give a formula $\frac{t}{t_*} = b\,N^a$ establishing a power dependence of the calculation time of the received signs. Parameters a and b have constant values for this computer. We will say that two calculations (N_1,t_1) and (N_2,t_2) are similar to each other if they have the same a and b. Thus, the last formula can be characterized as an experimental law of similarity of calculations. The formula allows to estimate calculation time for a large number of signs of the number π after rather small number of such signs is found.

Today the number π is calculated with the astronomical precision: 13 000 000 000 000 decimal signs (2014). Now the hardly foreseeable set of formulae, mathematical and physical facts are connected with this number. Their quantity continues to grow promptly.

For comparison, we will consider the sequence of the famous mathematical constants, according to their complexity growth of their calculation: 1, τ, e, π, $\alpha\,\&\,\delta$. Here $\alpha\,\&\,\delta$ are hardly computable constants of Feigenbaum.

6. The sine function

To be exact a sinusoid $y = a\sin(bx+c)$, is the simplest and most important function among all periodic functions. It was clear for Galileo, Newton, Fourier, Maxwell, Rayleigh and de Broglie.

Galileo Galilei (Dialogue Concerning the Two Chief World Systems – Ptolemaic and Copernican, The first day [1]) has formulated three simple curves: straight line, circle and spiral (helix). In fact, the simple curve is the same thing as a spiral, and the line and the circle – special cases. There is only one smooth connected regular curve in the three-dimensional space. This is the spiral. It is the regular curve in the sense that each its point is equally surrounded by the other points of the curve. According to differential geometry, curvature and torsion at each point of the spiral are constants. The projection of the spiral on a plane parallel to an axis of symmetry of the spiral gives a sinusoid.

Isaac Newton (Mathematical Principles of Natural Philosophy, Proposition XLVII, Theorem XXXVII [2]) asserted that: "If impulses are propagated trough a fluid, the several particles of the fluid, going and returning with the shortest reciprocal motion, are always accelerated or retarded according to the law of the oscillation pendulum". In the modern language atoms and molecules of liquid perform harmonic or sinusoidal oscillations. Another important facts relating to the sine function can be found in the literature [3-5].

7. The new formula

Which are given here for the sine and the number π will be useful in the theory of harmonic oscillations and waves.

References

[1] Galilei G 1632 *Dialogo sopra i due massimi sistemi del mondo tolemaico e copernicano* (Fiorenza)
[2] Newton I 1687 *Philosophiae naturalis principia mathematica* (Londini)
[3] Cauchy A L 1821 *Analise algébrique* (Paris)
[4] Fourier J 1822 *Théorie analytique de la chaleur* (Paris)
[5] Baron Rayleigh 1877, 1878 *The theory of sound* vol I, II (London: Macmillan and co.)

Journal of Physics: Conference Series **731** (2016) 012002 doi:10.1088/1742-6596/731/1/012002

Research into the process of impingement of two plane jets of an ideal fluid with free boundaries

V D Baskakov, K A Karnaukhov
Department of Physics, Bauman Moscow State Technical University, Moscow, Russia
E-mail: karnaukhov93@mail.ru

Abstract. The problem of finding parameters stationary jets, outgoing from a place of impact of two incoming flat ideal jets with free boundaries and possessing the identical speed, but various width, has no decision. Various models are widely used for a conclusion of the missing equation now, but they lead to contradictory results. The new model is offered. Adequacy to the offered model was checked by comparison results with data of the numerical calculations in ANSYS AUTODYN. The dependence approximating results of numerical calculations is developed to increase accuracy in calculations of angular provision of the internal outgoing jet.First of all, the executed researches are interesting to experts, who works on behavior low-value technological errors in shaped charges.

In one of the classic problems of theory of perfect fluid jets (Fig. 1) it is required to calculate the parameters δ_3, δ_4, β, γ of the outgoing jets provided that the values of the parameters α, δ_1 and δ_2 are specified. The solution to this problem is important for designing devices of inject automatics and development of technologies for the manufacture of cones shaped charges.

The problem has no unequivocal solution, for to be able to define 4 unknowns one may write only 3 separate equations based on the mass and momentum conservation laws.

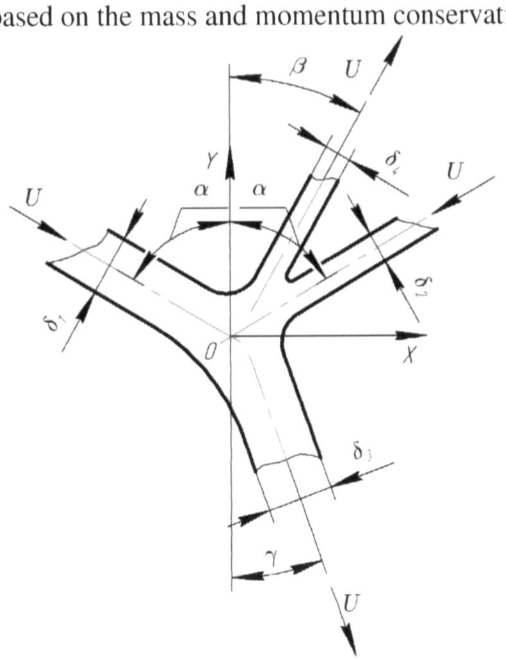

Figure 1. Impingement of two plane jets of a perfect fluid with free boundaries

δ_1, δ_2 – the width of the impinging jets ($\delta_1 > \delta_2$); 2α – the collapse angle of the impinging jets ($2\alpha < \pi$); δ_3, γ – the width and the deflection angle of the outer jet; δ_4, β – the width and the deflection angle of the inner jet ($\beta < \alpha$)

$$\begin{aligned} \bar{\delta}_3 + \bar{\delta}_4 &= 2, \\ 2cos\alpha &= \bar{\delta}_3 cos\gamma - \bar{\delta}_4 cos\beta, \\ 2\Delta\bar{\delta} sin\alpha &= \bar{\delta}_3 sin\gamma + \bar{\delta}_4 sin\beta, \end{aligned} \quad (1)$$

where $\bar{\delta}_3 = \delta_3/\delta_0$, $\bar{\delta}_4 = \delta_4/\delta_0$, $\Delta\bar{\delta} = \Delta\delta/\delta_0$, $\delta_0 = (\delta_1 + \delta_2)/2$, $\Delta\delta = (\delta_1 - \delta_2)/2$ – the deflection of the width of the impinging jets from the average value of δ_0.

The problem (1) may be solved without introducing any additional conditions only in two specific cases: if the impinging jets have equal widths $\delta_1 = \delta_2 = \delta_0$ or if the jets of different widths impinge at $\alpha = \pi/2$.

By calculating kinetic energy in a system of interacting jets, A. Palanini has established that it reaches its minimum if the directions of the outgoing jets are opposite:

$$\beta = -\gamma. \quad (2)$$

From solving (1) and (2) combined, we may derive the following:

$$\beta = -\gamma = -arctg(\Delta\bar{\delta} tg\alpha), \quad (3)$$

$$\bar{\delta}_3 = 1 + cos\alpha\sqrt{1 + \Delta\bar{\delta}^2 tg^2\alpha}, \ \ \bar{\delta}_4 = 1 - cos\alpha\sqrt{1 + \Delta\bar{\delta}^2 tg^2\alpha}.$$

The condition (2) contradicts the results of later researches [1 – 4]. The model developed on the basis of the concept of elastic/non-elastic interactions of jet flows is regarded as a more advanced. It states that the angles at which the outgoing jets tilt towards the OY axis are equal [1]:

$$\beta = \gamma. \quad (4)$$

The combined solving of (1) and (4) determines that:

$$\beta = \gamma = arcsin(\Delta\bar{\delta} sin\alpha), \quad (5)$$

$$\bar{\delta}_3 = 1 + cos\alpha, \ \ \bar{\delta}_4 = 1 - cos\alpha.$$

While calculating angle deflections and sizes of outgoing jets. J.P. Curtis proceeded from two basic propositions [2]: a slug is formed in the region of the point 0 of the impinging jets (Fig. 1); the transverse sizes of the impinging jets are to each other as δ_1 is to δ_2; it is also assumed that β and γ are of small value.

$$\beta = \Delta\bar{\delta} ctg\frac{\alpha}{2}, \ \ \gamma = \Delta\bar{\delta} tg\frac{\alpha}{2}. \quad (6)$$

At the same time, the widths of outgoing jets are defined by the relation from the model (5).

One of most important conclusions that may be drawn from the analysis of the models determining the parameters of outgoing jets is as follows: the widths δ_4, δ_3 either conform to the symmetric case (5) [1, 2] or approximate it. In this connection, one may use as one's basic proposition allowing one to solve the system of equations (1) the assumption that the widths of outgoing jets is only determined by the mean width δ_0 of the impinging jets and the angle between them and does not depend on the difference $\Delta\delta$ in the width of the impinging jets. Thus, solving the system of equations (1) and assuming that angles β and γ are low, i. e. accepting that $sin\beta \approx \beta$, $sin\gamma \approx \gamma$, $cos\beta \approx 1 - \beta^2/2$, $cos\gamma \approx 1 - \gamma^2/2$, yields the following:

$$\beta = 2\Delta\bar{\delta} f_1(\alpha), \ \ \gamma = 2\Delta\bar{\delta} f_2(\alpha), \quad (7)$$

where $f_1(\alpha) = \frac{1}{1+tg\frac{\alpha}{2}}$, $f_2(\alpha) = \frac{tg\frac{\alpha}{2}}{1+tg\frac{\alpha}{2}}$.

Dependencies (7) and the already mentioned model (6) ensure that the condition $\beta > \gamma$ is fulfilled.

It is significant that if $\alpha = 90^0$ the obtained relations (7) and the models (5), (6) yield identical value of the angles of outgoing jets, which also agrees with the exact solution (3) of the problem (1) provided that $\Delta\bar{\delta} \ll 1$.

In order to compare and analyse the models (3, 5, 6, 7), the parameters of outgoing jets were determined in numerical system ANSYS AUTODYN in dynamic view in the Euler coordinate system taking into consideration the compressibility the medium according to the recommendations provided in the work [4].

The following system of equations was solved numerically:

1) Equations of motion

$$\frac{\partial u}{\partial t} + \frac{\partial u}{\partial x}u + \frac{\partial u}{\partial y}v + \frac{1}{\rho}\frac{\partial p}{\partial x} = 0,$$
$$\frac{\partial u}{\partial t} + \frac{\partial v}{\partial x}u + \frac{\partial v}{\partial y}v + \frac{1}{\rho}\frac{\partial p}{\partial y} = 0.$$

2) Continuity equations

$$\frac{\partial \rho}{\partial t} + \frac{\partial \rho}{\partial x}u + \frac{\partial \rho}{\partial y}v + \rho(\frac{\partial u}{\partial x} + \frac{\partial v}{\partial y}) = 0,$$

where u, v – the components of the velocity vector received as functions of the coordinates of points (x, y) and time t, p – pressure, ρ – density.

The speed of blast wave U_s in a fluid was defined with this dependency [4]:

$$U_s = c_0 + s \cdot u_p,$$

where c_0 – the speed of sound, u_p – the speed of stream of the fluid, s - parameter.

During the performance of calculations the value was $s = 5$, in accordance with the recommendations in the work [4].

The design field of the problem included three interrelated sub-fields unfilled with medium (white areas in Fig. 2, 3). The lateral rhombus-shaped sub-fields were intended for impinging jets, the central – for outgoing ones. The impinging jets are not restricted in length and flow into design sub-fields uninterruptedly. The speed of stream of the impinging jets was $U = 3$ m/s, the density - $\rho_0 = 1000$ kg/m^3, the speed of sound - $c_0 = 1500$ m/s. The condition $U \ll c_0$ is essential for the problem to approximate the model of an uncompressible fluid. The *Hydro (Pmin)* model was accepted as a failure model of fluid.

During the course of calculation, the following parameters varied: $k = \delta_2/\delta_1 = (1-\Delta\bar{\delta})/(1+\Delta\bar{\delta})$, which assumed the following values/magnitudes $k = 0{,}5$; $k = 0{,}667$; $k = 0{,}75$; $k = 0{,}85$; the angle between the incoming jets 2α, which varied within the range of 90^0 to 180^0 with a 10^0 step. Numerical calculations with different angles of the impingement of jets differed in the tilt of the lateral sub-fields (Fig. 2):

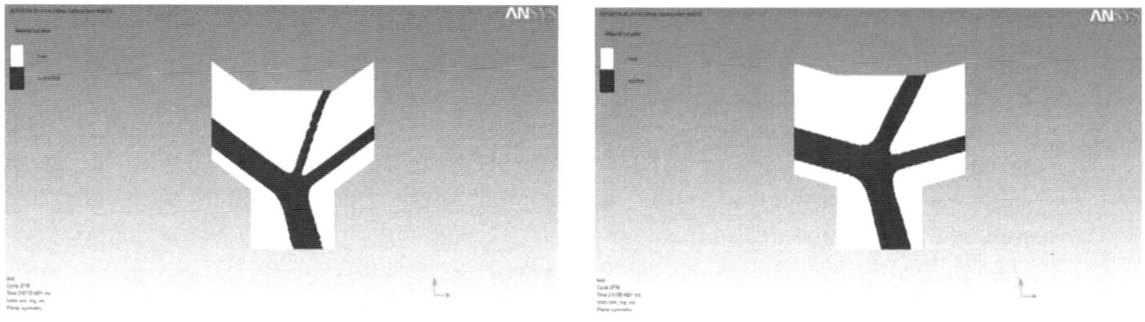

a) b)

Figure 2. Impingement of two plane jets with $k = 0{,}5$: a) $\alpha = 55^0$, b) $\alpha = 75^0$

In the course of performing the calculations, it was observed that from the moment $t \approx 25$ ms the non-stationary stage of the jet interaction transforms into a stationary one and the jet flow configuration doesn't change henceforth (Fig. 3).

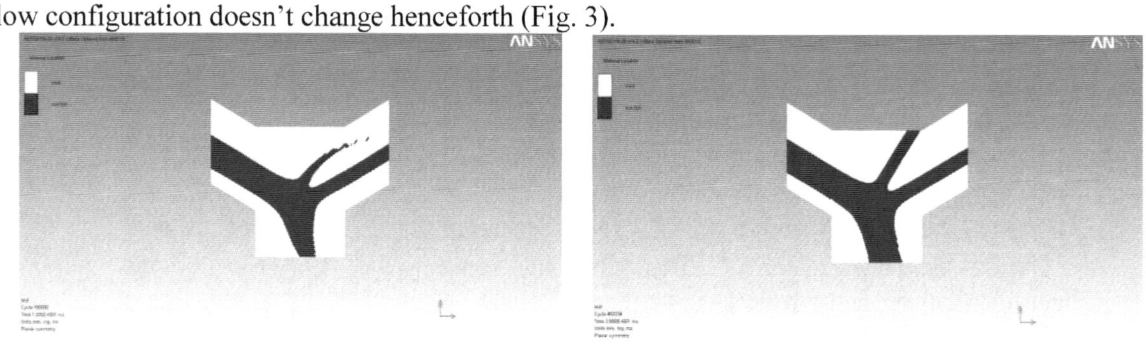

a) b)

Figure 3. Impingement of two plane jets with $\alpha = 60^0$ and $k = 0{,}5$: a) non-stationary stage ($t = 10$ ms), b) stationary stage ($t = 25$ ms)

The comparison of the results of numerical calculations of the outgoing jets' width with the model of invariability in size of the outer δ_3 and the inner δ_4 of the jets during the transition from a symmetric to an asymmetric impingement that has been proposed in the article has shown that the assumption made in this article that it is possible to use the dependencies (2) to estimate the width of outgoing jets is acceptable not only if $\Delta\bar{\delta} \ll 1$, but also if there are substantial distinctions in the overall dimensions of the impinging jets.

A comparison has been drawn between the results of the numerical calculations of the angle deflections of the inner $\beta = \beta(\alpha)$ of the outgoing jet (curve 2 in Fig. 4) and calculations according to the developed model (7) and the already known models (5), (6) (curves 1, 3, 4 in Fig. 4 respectively). The calculations for the outer jet $\gamma = \gamma(\alpha)$ agree among themselves and are therefore not adduced in the article.

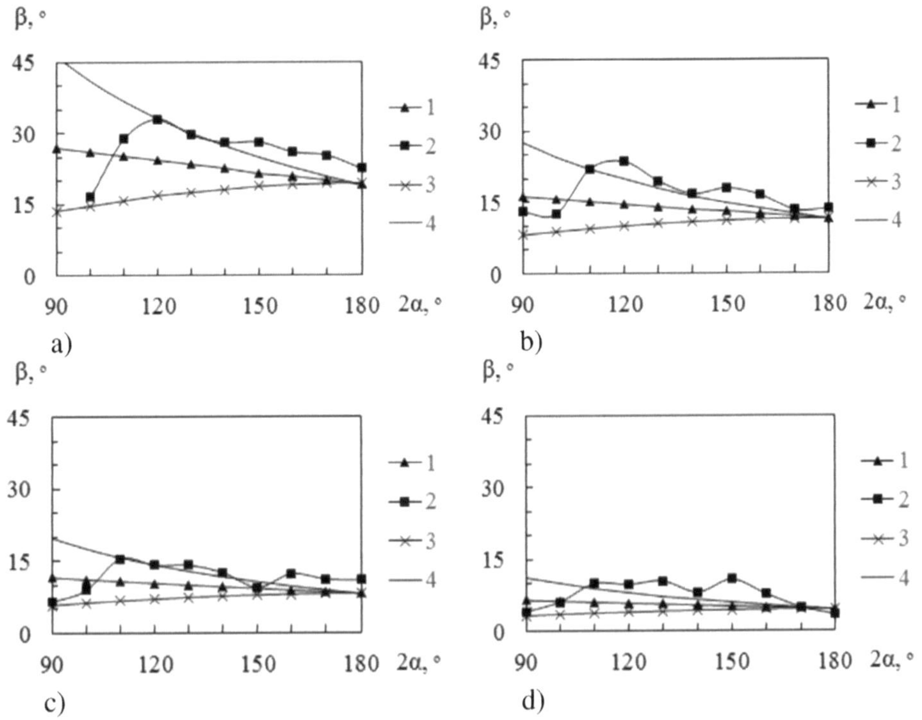

Figure 4. Angle deflection β of the inner jet depending on the angle 2α between the impinging jets:
a) k = 0,5; b) k = 0,667; c) k = 0,75; d) k = 0,85

The analysis of Fig. 4 reveals that if $k = 0{,}85$ then the results of the calculations according to different models are quite similar. However, if the difference in the widths of the impinging jets is substantial (for example, in case $k = 0{,}5$), one may observe not only considerable discrepancy in calculations, but also contradictory tendencies in the character of the change of curves $\beta = \beta(\alpha)$ in different models (for example, graphs 3, 4 in Fig. 4a have different signs of curvature).

A special attention should be paid to errors in numerical calculations during the estimation of the angle β. For instance, if $2\alpha = 180^0$ then numerical calculations should agree with the dependency (3), as well as the models (5), (6), (7) (Fig. 4). The observable discrepancies of the calculation results allow one to assume random errors connected with inaccuracies in the measurement of the angle β in design graphs of stationary flow of plane jets (Fig. 2, 3b). To compensate the errors, the following approach may be adopted. Given that the developed (7) and the already known (5), (6) analytical dependencies indicate proportionality between β and $\Delta\bar{\delta}$, it may be accepted that proportionality must also be maintained in numerical calculations, i. e.

$$\beta = \Delta\bar{\delta} f(\alpha). \qquad (10)$$

From processing the results of numerical calculations, the following function $f(\alpha)$ is derived.

$$f(\alpha) = 6(2\alpha)^{10} e^{-4{,}43(2\alpha)} \qquad (11)$$

In the equations (10), (11), the angles α and β are measured in radians.

The comparison of the results of numerical calculations and calculations through dependencies (10), (11) illustrates their agreement if the condition $k \geq 0{,}667$ ($\Delta\bar{\delta} \leq 0{,}2$) is fulfilled.

Such discrepancy $\Delta\bar{\delta}$ in the width of impinging jets is quite acceptable for solving practical problems, for example, appearing in the process of estimation of how low-value/magnitude errors of shaped charges affect the angle deflection of the forming inner jet. Besides, the results of the conducted research may be used for developing fluidics devices, as well as for setting conditions explosion welding of metals.

References

[1] Kinelovskii S A 1994 Concept of elastic–inelastic interactions of jet flows and collision of jets of an ideal incompressible fluid. *Combustion, Explosion, and Shock Waves* **30** iss. 3, pp 331–339
[2] Curtis J P 1998 Asymmetric Formation of Shaped Charge Jets. *Proceedings of the 17th International Symposium on Ballistics, Midrand, South Africa* **2** pp 405–412
[3] Kinelovskii S A and Sokolov A V 1986 Nonsymmetric collision of plane jets of an ideal incompressible fluid. *Journal of applied mechanics and technical physics* **27** iss. 1 pp 47–50
[4] Miller S, Ceder R 2001 The indeterminacy of the outgoing flow of two impinging asymmetric jets. *Proceedings of the 19th International Symposium on Ballistics, Interlaken, Switzerland* pp 843–850

Space-time inversion and its consequences

M Chelnokov
Department of Physics, Bauman Moscow State Technical University, Moscow, Russia
E-mail: l-chelnok@yandex.ru

Abstract. The article discusses some new aspects of both the inversion of space and the inversion of time. It is shown that behind the mirror is not symmetric to in front of the mirror. It, in its turn, leads to nonconservation of spatial parity. The same situation takes place in the combined CP-parity. Further, the article shows that from the point of view of different reference systems of the Universe (from the point of view of different galaxies or accumulation of galaxies) time flows not just differently, and, in some cases, in the opposite directions. It leads to major changes in the picture of the Universe. In particular, the concept of the age of the Universe loses its meaning, serious doubts about the idea of the Big Bang and so on.
Keywords: mirror symmetry, true vectors, pseudovector, weak interaction, inversion, spatial parity, the Universe, events, time chaos.

1. Spatial inversion

Mirror symmetry refers to discrete symmetries and, as it is known, it is the basis of the law of conservation of P-parity – spatial parity. The point of view was considered and remained obvious until 1956. In the year of 1956 T.Lee and C.Yang in their famous article [1] carried out a systematic analysis of the law of conservation of P-parity in the processes with elementary particles and stated nonconservation of parity in the process caused by the weak interaction.

In 1957 group of physicists (Ambler, Hayward, Hopps and Hudson) under the leadership of C.Wu carried out a brilliant experiment [2] which showed that P-parity is not conserved in the β - decay of cobalt $_{27}Co^{60}$ caused by the weak interaction.

For the explanation of this phenomenon two hypotheses were suggested: the hypothesis of mirror asymmetry of the three-dimensional Euclidean space, and the hypothesis of the combined CP-parity in which the particles of the ordinary world are replaced by antiparticles behind the mirror. The second hypothesis was proposed by Wigner, Lee, Yang and Landau [3, 4].

The first hypothesis was not developed and it was almost forgotten. But the second hypothesis received the right for existence. But, firstly, it is just a hypothesis that can not be tested directly, as it is hardly possible to imagine how to detect antiparticles behind the mirror. This hypothesis seems to be initially confirmed, but only as it saves mirror symmetry.
Secondly, even the salvation proved to be illusory and of short duration, as 7 years later, in 1964, the experiment of Christensen and his colleagues [5] discovered nonconservation of CP-parity in the decays of kaons. The problem has been discussed and is being discussed in many sources (see, for examples, [6-10]), but no satisfactory conventional solution has been found up till now.

Now let us turn to our point of view. We show that the usual three-dimensional space, in general, does not possess mirror symmetry with respect to the conversion, and this very fact explains the nonconservation of spatial parity.

Generally, when considering the mathematical formalism which is associated with the concept of parity the mirror symmetry is mentioned in passing, and we go straight to the spatial inversion, which is equivalent to the reflection in three mutually perpendicular mirrors (see, for examples, [11, 12]). Thus, in our opinion some essential points are lost. We will discuss the mirror reflection in one mirror in more detail. We will call it a single mirror transformation.

First of all, let us discuss the mirror transformation component of true (polar) and pseudovectors (axial vector) which are perpendicular and parallel to the surface of the mirror. We mark true vector as A, pseudovector as B, the point of application of the vectors A and B as 0_A and 0_B. Strokes will be used to mark the same values after the mirror transformation. The pseudovector which is perpendicular to the surface of the mirror, in the mirror transformation does not change the direction. The pseudovector which is parallel to the surface of the mirror changes its direction for the reversed in the mirror transformation. For the true vector the situation is reversed. Let us mark the component of the vector which is parallel to the surface of the mirror as index 1 and the perpendicular component is index 2. Thus, we have the following laws of the mirror transformation:

$$A_1' = A_1 \; ; A_2' = -A_2 \; ; B_1' = -B_1 \; ; B_2' = B_2.$$

As we can see, in terms of the mirror transformation both true vectors and pseudovectors possess equal rights in case we swap their perpendicular and parallel components (Fig. 1).

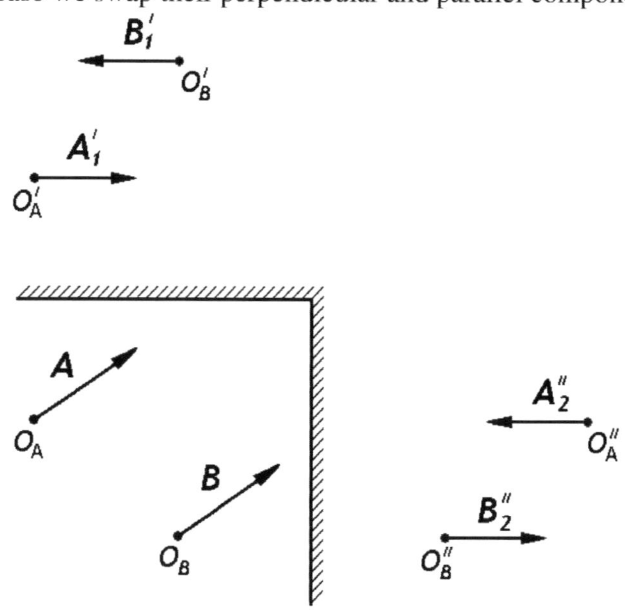

Figure 1. Mirror transformation of the parallel and perpendicular components of the true vector and pseudovector

If we consider the transformation of the perpendicular and parallel components of the true vector and the pseudovector and then add them up, we get the following picture (Fig. 2).

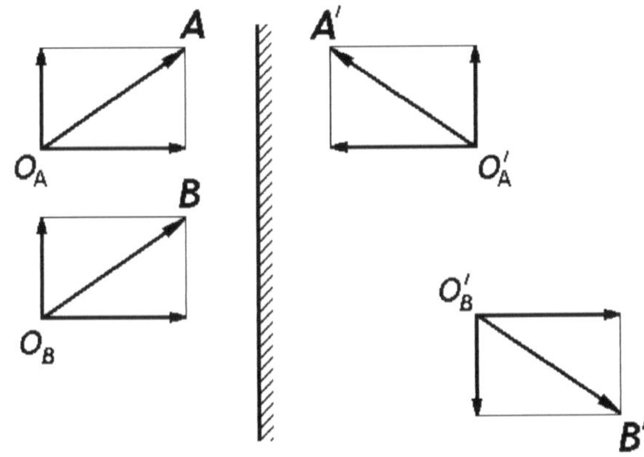

Figure 2. Mirror transformation of the true vector and pseudovector

Thus, if the real experimental situation is depicted by a set of true vectors and pseudovectors thus the mirror transformation of the whole picture is not symmetric.

Now let us consider the operation of space inversion: x → – x, y → – y, z → – z. Otherwise it can be interpreted as mirror reflection in the three mutually perpendicular mirrors (the order of such reflections is not important, all the options lead to the same result). The transformation of true vectors and pseudovectors under spatial inversion is shown in Fig. 3.

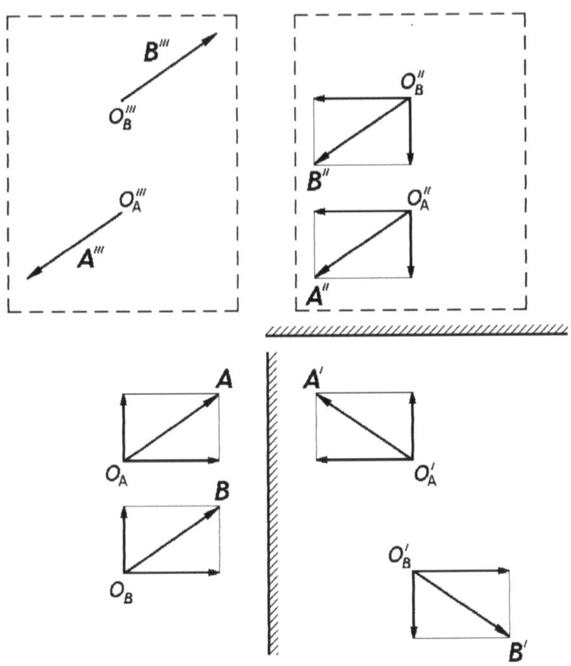

Figure 3. Spatial inversion of the true vector and pseudovector

Here are some explanations. The lower left part of the figure is our usual space, in which there is a true vector and pseudovector. The bottom right of the figure is the picture after the first mirror transformation. The upper right part of the figure is the picture after the second mirror transformation in the mirror, which is perpendicular to the first one.

In this part the dotted line depicts the third mirror, which is perpendicular to the first two mirrors. This mirror is parallel to the plane of the figure, but it lies under this plane.
In the upper left part of the figure shows the picture after the third mirror transformation in dotted mirror. This picture is also parallel to the figure plane, but lies below it, i.e. under the third mirror.

As we see, if the real experimental situation described by a set of true vectors and pseudovectors, then the initial and the final picture are not symmetric with respect to each other.

Thus, both single mirror transformation and spatial inversion in the presence of a set of true vectors and pseudovectors don't possess any symmetry.

Thus, the question «Why is P-parity conserved in a particular process?» is more assential than the question «Why is not P-parity conserved?». From this point of view the search for the nonconservation of P-parity in the process with strong and electromagnetic interactions makes sense. By the way, if the weak and electromagnetic interactions have already been combined into the electroweak, then why does not the nonconservation of P-parity in the electromagnetic interaction take place, if it exists in the weak one?

By the way, the introduction of combined CP-parity from the point of view of the considered picture does not help the situation and, as we know, CP-parity is notconserved in the decay of kaons.

2. Inversion of time

We now proceed to the consideration of the inversion of time. Until now it was considered that we can speak of the single direction of time in the Universe on the whole – from the past to the future. At present, it is well known that in the frame of the special theory of relativity (STR) time has different «speed» in different inertial reference systems (IRS) and can even «move» in opposite directions, and in the framework of the general theory of relativity, time slows down near gravitating bodies.

However, it did not prevent us from considering the flow of time on the whole, for example, the lifetime of the Universe since the Big Bang is estimated at 13 billion years.
Being entirely consistent and while analyzing the time in the framework of SRT, our idea of the single direction of time in the Universe loses its sense as, for example, the idea of the absolute simultaneity.

Besides it turns out that the known operation of inversion (both spatial and inversion of time) under some conditions equivalent to the operation of the transition from one IRS to another within SRT.

We introduce the following reasonable, in our view, definition of different direction of time:

Suppose that in some IRS K the event 2 occurs after the event 1, and IRS K′ event 1 occurs after event 2. So we can say that in these two IRS time flows in the opposite directions.
In both of these systems, time flows from the past to the future, but what is the future for IRS K turns out to be the past for IRS K′, and the past for IRS K appears to be the future for IRS K′.

Suppose there are two events: Event 1 and Event 2 (or 3). Let us consider them from the point of view of IRS K and IRS K′ (Fig. 4). Reference system K′ moves relative to the reference

Journal of Physics: Conference Series **731** (2016) 012003 doi:10.1088/1742-6596/731/1/012003

system K along coincident axes x and x' to the right with velocity V, or, what is the same, the reference system K moves relative to the reference system K′ to the left with velocity -V.

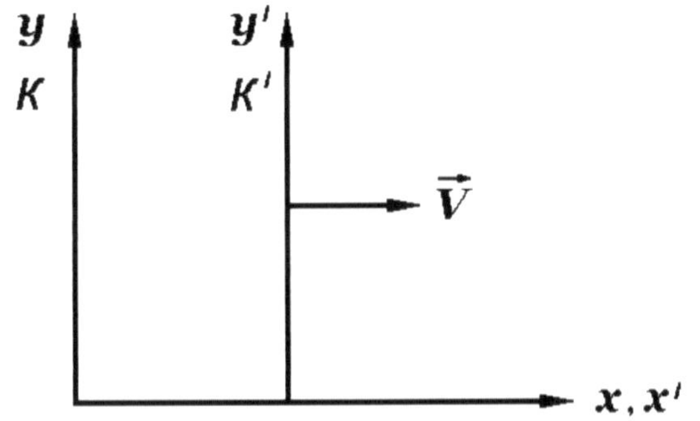

Figure 4. Two inertial reference systems in SRT

As it is known [13, 14], in the SRT have the following formulas of transformation of periods of time.

Transition IRS K′ → IRS K:

$$c\Delta t = c(t_2 - t_1) = \gamma[c(t_2{}' - t_1{}') + \beta(x_2{}' - x_1{}')]. \qquad (1)$$

Transition IRS K → IRS K′:

$$c\Delta t' = c(t_2{}' - t_1{}') = \gamma[c(t_2 - t_1) - \beta(x_2 - x_1)]. \qquad (2)$$

Here we introduce the usual symbols:

c – velocity of light;

$$\beta = \frac{V}{c},$$

$$\gamma = \frac{1}{\sqrt{1-\beta^2}}.$$

Primed quantities refer to the IRS K′ and the unprimed ones – to IRS K.

We begin with the formula (1). Let in IRS K′ event 2 occurent 1, so $t_2{}' > t_1{}'$ or $t_2{}' - t_1{}' > 0$. Then for the inversion of time it is necessary that $t_2 - t_1 < 0$, and then the second component of sum of the right-hand part of the formula (1) must be negative, so $x_2{}' - x_1{}' < 0$ and, in addition, the inequality must be carried out:

$$\beta(x_1{}' - x_2{}') > c(t_2{}' - t_1{}'). \qquad (3)$$

Let us consider the value v with the dimension of velocity:

$$v = \frac{x_1' - x_2{}'}{t_2{}' - t_1{}'} > \frac{c}{\beta}. \qquad (4)$$

As you can see, this value must be greater than the velocity of light, which is impossible, and therefore inversion of time is only possible for the two events which are not connected by the cause-and-effect relation (the interval between these two events is space-like).

Now let us take formula (2). Let in IRS K event 2 occurs after event 1, so $t_2 > t_1$ or $t_2 - t_1 > 0$. Then the inversion of time it is necessary $t_2{}' - t_1{}' < 0$, and the second component of sum of formula (2) must be negative, so $x_2 - x_1 > 0$ and, moreover, the inequality must be carried out:

$$\beta(x_2 - x_1) > c(t_2 - t_1). \qquad (5)$$

Let as consider, as the previous case, the value v with the dimension of speed:

$$v = \frac{x_2 - x_1}{t_2 - t_1} > \frac{c}{\beta}. \qquad (6)$$

This value is again greater than the speed of light, and therefore the inversion of time is only possible for the two events which are not connected by the cause-and-effect relation (the interval between these two events is space-like).

Let us give the geometric interpretation of the phenomena from the point of view of Minkowski world. Let us consider the transition IRS K′ → IRS K (Fig. 5). In the initial IRS K′ first event 1 occurs and then – event 2 (or 3) occurs. The case where the inequality (3) is not satisfied, corresponds to the point 2. Here and in IRS K, as in the original IRS K′ event 2 occurs after the event 1, i.e. no inversion of time.

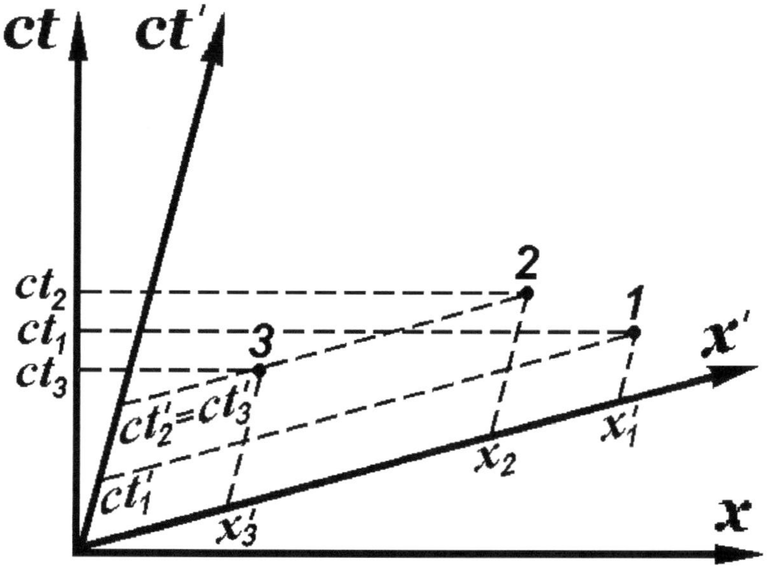

Figure 5. The picture of inversion of time in Minkowski world. The transition from IRS K′ to IRS K

The case when the inequality (3) is satisfied corresponds to point 3. Here in IRS K event 3 occurs, as opposed to the IRS K′, before event 1, so the inversion of time occurs. (Or if the inequality (3) becomes the equality, the event 3 occurs simultaneously with event 1).
Now let us consider the transition IRS K → IRS K′ (Fig. 6). In the initial IRS K event 1 occurs first, and then event 2 (or 3) occurs. The case, when the inequality (5) is not satisfied, corresponds to the point 2. Here in IRS K′, as in the initial IRS K event 2 occurs after event 1, so there is no inversion of time.

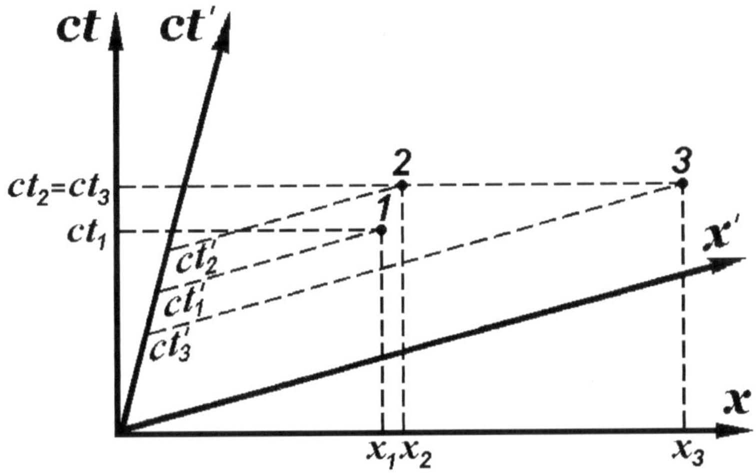

Figure 6. The picture of inversion of time in Minkowski world.
The transition from IRS K to IRS K′.

The case, when the inequality (5) is satisfied, corresponds to the point 3. Here, in IRS K′ event 3 occurs, unlike IRS K, before the event 1, i.e. there is the inversion of time. (Or, if inequality (5) becomes equation, the event 3 occurs simultaneously with event 1).

Thus, the conditions under which there is the inversion of time, or, in other words, mutually opposite directions of the flow of time for the two of IRS are connected with the following three factors:

1. Relative velocity of IRS.
2. Spatial distance between two points.
3. Period of time.

Let us consider all possible pairs of the IRS connected with real bodies of the Universe. In some of these pairs (and in some pairs of points) time flows in the same direction, and in some of them time flows in the opposite direction. Thus, the idea of the single direction of time in the Universe loses its sense. Past and future are relative. In the Universe temporal chaos reigns. The question of the age of the Universe loses sense too.

3. Evaluation of the order of values and experimental scheme

Now let us estimate the order values describing different directions of time in the Universe. We use formula (3) in this case. The typical distance beginning from which the Universe can be considered homogeneous, and from which Friedman dynamic solution considered true about 100 Mps. Let us take 100 Mps as the distance between two points in which two events occur. Let us the relative velocity of the two reference systems (for example, of two galaxies) be 0.1 light speed.

Then, for the inversion of time the interval between the two events in one reference system must not exceed 30 million years. Thus, from the point of view of the reference system which is connected with one of the galaxies, event 1 occurs 30 million years earlier than event 2, then from point of view of the reference system which is connected with the other galaxy, event 1 occurs 30 million years later then event 2. Thus, we cannot speak of any arrow of time.

It is not difficult to select such parameters in the Universe under which inversion time difference exceeds the age of the Universe (13 billion years).

As it is known, the radius of the Universe is about 12000 Mps. Let us take 10000 Mps as the distance between the two points and relative velocity of the reference systems (in which the two events occur) about the velocity of light. Then the difference of time is equal to 30 billion years. This value proves that the notion of the age of the Universe loses its sense.

We would like to stress that the given here calculations of the inversion of time are based on the special theory of relativity, and therefore they are obvious. Nevertheless, carrying out of an experiment confirming the inversion of time may be of some interest to us. The scheme of such an experiment could be next.

Let the primed IRS K′ is connected with the Earth, and the unprimed IRS K - with two satellites, which have the zero velocity relative to each other and they are at some distance from each other. The clocks on these satellites are synchronized. Signals are sent from the satellites – at first from the one and then – from the other one. On Earth, these signals are received in reverse order.

Let us calculate the parameters of such an experiment. In order to get the inversion of time the second member of sum in equation (2) must be negative and grater in module than the first member of sum.

Take the velocity of the satellites relative to the Earth V = 10 Km/s = 10^4 m/s $\Delta x = x_2 - x_1 = 10^3$ Km $= 10^6\ m$ – about 2,5% of the length Earth's equator.

Then the formula (5):

$$\Delta t = t_2 - t_1 < \frac{\beta}{c}(x_2 - x_1) \approx 10^{-7}\text{s} = 100\text{ ns}.$$

Take $\Delta t = 0{,}3 \cdot 10^{-7}\text{s} = 30$ ns. Then the interval of time between the signals recorded on the Earth, is calculated by the formula (2) and its numerical value is $\Delta t' = 0{,}8 \cdot 10^{-7}\text{s} = 80$ ns. The experiment seems to be feasible under these parameters nowadays.

4. Picture of the Universe

Let us connect each galaxy (or galaxy cluster) with their inertial reference system (IRS). Thus, let us imagine the whole Universe as the combination of such IRS. However, the question if the system is inertial or not, arises. However, taking into account the gravitational interaction, there are no inertial systems in the real Universe. Thus, we can introduce some instantly accompanying system of reference. What is considered to be an inertial standard system in which the acceleration is zero?

Apparently, two variants can be offered. Firstly, this is the reference system with regard to which cosmic microwave background radiation is strictly isotropic. Secondly, this is the reference system with regard to which dark matter generally rests. Are such systems identical to each other? Apparently, this question has no answer.

As we have already seen, the theory of relativity implies that the time in different inertial reference systems flows differently, including possibly in opposite directions. In that case the concept of the age of the Universe as a whole has no sense. This concept is only maintained for each IRS separately. How can such a Universe be described? It is pertinent here to draw analogy with the concept of temperature. As you know, temperature makes sense only for equilibrium systems, and it loses its sense for uneven-equilibrium systems.

The description of the uneven-equilibrium systems is associated with different distribution functions, which are more diverse and complex for them than for the equilibrium ones. Apparently, the Universe, in which from the point of view of different IRS, time flows differently, should be described with the help of using some distribution functions which have to be introduced.

Now the question is if this or that event and, in particular, the Big Bang took place in the Universe. In the light of the said above, it seems to us that there may be at least four possible answers to this question. Let us consider them.

1. Yes, this event (we will speak, for definiteness, of the Big Bang) took place. At present it is a generally accepted answer.
2. No, there wasn't.
3. The question does not make sense, since the concept of the age of the Universe makes no sense.
4. From the point of view of some IRS – there were, from the point of view of some others – there were not.

Thus, from the point of view of the second group of IRS two more subvariants are possible. The first subvariant: the Big Bang in the second group of IRS will take place in the future, the second subvariant: the Big Bang will never take place. By the way, from the point of view of those IRSs, in which the Big Bang took place, it happened 13-14 billion years ago, as it is generally accepted, but it took place at different time in different IRSs. So, the main conclusion derived from these arguments in the following: the Universe looks much different from the point of view of different reference systems. Thus, if the Big Bang did not take place, or the question does not make sense (the second and third versions of the answer), then all known fundamental phenomena (expansion of the Universe, the cosmic microwave background radiation, etc.) should be interpreted in a new way. Remember, that this situation is not unique, it has already taken place many times in the history of physics. For example, in the late nineteenth century, many phenomena were interpreted within the concept of the ether. Then the concept vanished, and all the phenomena remained, of course, but they began to be interpreted differently. If the fourth answer is true, then the situation becomes even more curious. From the point of view of the IRS in which we exist, the Big Bang probably took place. A large number of experimental facts fit into this scheme.

In general all of IRS (galaxies or their clusters) are divided into two groups. From the point of view of one of them the Big Bang took place, from the point of view of another – it didn't. The description of this strange Universe – is the goal for the future.

5. Conclusions

The main conclusions of this work are the following:

1. The inversion of time and spatial inversion (mirror reflection) to certain extent are equivalent to the transition from one IRS to another. Thus, it is possible to some extent to simulate inversion of time and mirror transformation.
2. Behind the mirror is not symmetric to in front of the mirror. It, in its turn, leads to nonconservation of spatial parity and CP-parity.
3. In the Universe in General there is no single direction of time. Call it a temporary chaos of the Universe. In connection with this question of the Big Bang remains open.
4. In the Universe on the whole is no single direction of time. Let us call it the chaos of time of Universe.

We believe this work can be developed further in the following directions:

1. The experimental search of the non-conservation parity in the processes caused by strong and electromagnetic interactions.

2. Development of methods for describing properties and behavior of big system (the Universe) in which time is chaotic but not dynamic.
3. Coordination of the introduced here concepts with the aggregate of all known in cosmology experimental data.
4. Analysis of real scheme of the possible experiment (and its setting), which confirms the introduced here concepts.

Acknowledgements
The work was supported by grants of RFFI№ 16-02-00488 A and № 16-08-00618 A.

References

[1]Lee T D and Yang C N1956Proposals to test spatial parity conservation in weak interactions. *The Physical Review* **104** no. 1 pp 254-258
[2]Wu C S, Ambler E, Hayward R W 1957 Experimental test of parity conservation in beta decay. *The Physical Review* **105** pp 1413-1415
[3]Wigner E P 1970 *Symmetries and Reflections Scientific essays* (Cambridge, Mass.: M.I.T. Press)
[4]Lee T D and Yang C N 1957 *New symmetry properties of elementary particles* (Moscow : World))
[5] Christensen J H and Cronin J W 1964 Evidence for the 2 pi decay of the K20 Meson. *Phys. Rev. Lett.* **13** pp 138-140
[6] Gibson W M and Pollard B R 1979 *Symmetry Principles in Elementary Particle Physics* (Cambridge, London, New York, Melbourne : Cambridge University Press) p 344
[7] Okun L B 1990 *Leptons and Quarks* (Moscow : Science) p 352
[8] Kane G 1987 *Modern Elementary Particle Physics* (Michigan : Addison-Wesley Publishing Company) p 360
[9] Chelnokov M B2010 On Spin of Fundamental Particles. *Herald of BMSTU. Natural Science* **3** (38) pp 22-34
[10] Chelnokov M B 2010 On Spin Projection of Fundamental Particles and Problem of Non-Conservation of CP-Parity. *Herald ob BMSTU. Natural Science* **4** (39) pp 73-85
[11] Bogolyubov N N and Shirkov D V1980 *Kvantovie polya. [Quantum Fields]* (Moscow : Nauka) p 320
[12] Bogolyubov N N and Shirkov D V 1984 *Vvedenie v teoryu kvantovih poley [Introduction in Theory of Quantum Fields]* (Moscow : Nauka) p 600
[13] Ugarov V A 1997 *The special theory of relativity* (Moscow : Nauka) p 384
[14] Fok V A 1955 *The theory of space, time and gravity* (Moscow : USSR) p 504
[15] Chelnokov M B2015 Asymmetry through the looking glass. *Chebyshev collection* **16** no. 1 pp 281-290

The models of cosmological inflation in the context of kinetic approximation

I Fomin
Assistant Professor, Physics Department, Bauman Moscow State Technical University, Moscow, Russia
E-mail: ingvor@inbox.ru

Abstract. In this work the building of models of cosmological inflation with approximate linear dependence of the scalar field kinetic energy on the state parameter is considered. The key parameters of cosmological perturbations are also calculated.

The kinetic approximation

The theory of the cosmological inflation successfully describes the condition of the accelerated expansion of the Universe evolution at early stages [1]. Also inflationary cosmology explains an origin of primary perturbations and predicts their spectrum.

According to the theory of inflation, the source of the primary perturbations is the quantum fluctuations of a scalar field [2]. These fluctuations had essential amplitudes in the scales of Planck length and they came nearer to scales of galaxies with almost same amplitudes during inflation. Thus, inflation connects the large-scale structure of the Universe with microscopic scales.

The resultant spectrum of the perturbations practically doesn't depend on private scenarios of inflation and has a universal form. It leads to unambiguous predictions for the CMB spectrum [3]

The theory of cosmological inflation successfully describes a condition of the accelerated expansion of the Universe at early stages of evolution [1, 2]. Also inflationary cosmology explains an origin of primary not uniformity and predicts their range. The mechanism of the initial inflationary scenario and the repeated accelerated expansion can be described in the assumption of existence of the scalar field or inflaton, defining stage of inflation and the field of quintessence which is a source of the observed accelerated expansion of the Universe.

According to the theory of inflation, the source of the primary perturbations is the quantum fluctuations a of a scalar field. These fluctuations had essential amplitudes in the scales of Planck length and they came nearer to scales of galaxies with almost same amplitudes during inflation. Thus, inflation connects the large-scale structure of the Universe with microscopic scales. The resultant spectrum of the perturbations practically doesn't depend on private scenarios of inflation and has a universal form. It leads to unambiguous predictions for the CMB spectrum.

Cosmological acceleration specifies that now in the Universe evenly distributed slowly changing space liquid with negative pressure, called by dark energy, dominates.

For the specification of various types of cosmic liquid the phenomenological ratio between a pressure p and an energy density ε, $p = w\varepsilon$ is usually used. Modern experiments [4] testify that the Universe is spatially flat and now the parameter of a condition of dark energy is $w = -1 \pm 0.1$.

Standard way of obtaining time-dependent the parameter of a state is inclusion of scalar fields in cosmological model. Under sufficiently general assumptions, within four-dimensional quintessential model with one scalar field $-1 < w < -1/3$ and phantom models of $w < -1$.

Journal of Physics: Conference Series **731** (2016) 012004 doi:10.1088/1742-6596/731/1/012004

Models of inflation are set by a type of effective potential $V = V(\phi)$. In this case, potential determines the behavior of a scalar field ϕ which comes down to $V(\phi)$ minimum. The end of inflation leads to violation of conditions of slow rolling, the field oscillates about a minimum and process of reheating begins. This process includes at once some various stages, such as the decay of inflation condensate, the birth of particles of standard model and their thermalization.

The equations of scalar field dynamics in flat Friedmann-Robertson-Walker Universe in units $8\pi G = c = 1$ are [3]

$$3H^2 = \frac{1}{2}\dot{\phi}^2 + V(\phi), \tag{1}$$

$$\ddot{\phi} + 3H\dot{\phi} = -\frac{dV(\phi)}{d\phi}, \tag{2}$$

$$\dot{H} = -\frac{1}{2}\dot{\phi}^2 . \tag{3}$$

From three equations of the scalar field dynamics are independent two only.

The building of consistent model of cosmological inflation requires performance of the following conditions: the existence of the stage of accelerated expansion which means $-1 < w < -1/3$, the stage of the accelerated expansion comes to the end with a reheating with the subsequent birth of photons, that is transition to a stage of radiation domination to which corresponds $w = 1/3$, the correspondence of received cosmological parameters to the observations.

The usual method for studying of inflationary dynamics is slow roll approximation, which means simplification of the equations of the scalar field dynamics by neglecting of kinetic energy [3].

Thus, in the case of the slow roll approximation $\frac{1}{2}\dot{\phi}^2 \approx 0$, that limits a form of potential and provides the existence of an inflationary stage.

The energy density and the pressure for a scalar field

$$\varepsilon = \frac{1}{2}\dot{\phi}^2 + V = X + V, \tag{4}$$

$$p = \frac{1}{2}\dot{\phi}^2 - V = X - V, \tag{5}$$

where X is kinetic energy of the scalar field ϕ.

The state parameter for the slow roll approximation

$$w = \frac{\frac{1}{2}\dot{\phi} - V}{\frac{1}{2}\dot{\phi} + V} \approx -1 \tag{6}$$

and Hubble parameter $H \approx const$.

It is also possible to receive the exact solutions of the equations (1) – (3) by means of various methods for arbitrary potentials [5-17]. In case of exact solutions, for which restrictions for a form of potential are absent [9], the proof of existence of a stage of the accelerated expansion and an exit from inflation, proceeding from a type of a scale factor or the state parameter is required.

Journal of Physics: Conference Series **731** (2016) 012004 doi:10.1088/1742-6596/731/1/012004

In this work the possibility of solutions of the scalar field dynamical equations without the restrictions for a form of potential and kinetic energy with exit from the inflationary stage is considered.

We will consider kinetic energy X of the scalar field as linear function of the state parameter so that at $w=-1$ the kinetic energy $X=0$, which corresponds to the accelerated expansion caused by the cosmological constant

$$X=\beta(w+1). \tag{7}$$

The state parameter

$$w=\frac{X-V}{X+V}. \tag{8}$$

Substituting (8) in the equation (7), taking into account the equation (1), we will receive $3H^2=X+V=2\beta$, $H=const$.

For the phantom fields $\beta<0$ and $w=(X+V)/(X-V)$ what again leads to result $H=const$.

Therefore, determination of kinetic energy of the scalar field in the form of $X\approx\beta(w+1)$, which we will call *kinetic approximation*, provides the accelerated expansion $H\approx const$ and gives the method of solution of the scalar fields dynamical equations by the choice of the state parameter w.

One can rewrite the equations (1) and (3) in the other form

$$V=3H^2-X, \tag{9}$$

$$\dot{H}=V-3H^2, \tag{10}$$

$$\dot{\phi}=\pm\sqrt{2}\sqrt{3H^2-V}. \tag{11}$$

Substituting the expression for kinetic energy, we get

$$V=3H^2-\beta(w+1), \tag{12}$$

$$\dot{H}=-\beta(w+1), \tag{13}$$

$$\dot{\phi}=\pm\sqrt{2\beta(w+1)}. \tag{14}$$

Now we will consider various cases of the state parameter as the function of Hubble parameter and time.

The state parameter as the function of Hubble parameter

For the state parameter $w(H)=-1+\dfrac{AH^n}{\beta}$ we will receive the following solutions of the equations (12) – (14)

$$V(\phi)=\frac{A^{3/2}}{\sqrt{8}}(2-n)(\phi+B)^{2/(2-n)}\left(\frac{3}{8}(2-n)^2(\phi+B)^2-\left(\frac{A}{8}\right)^{n/2}(2-n)^n(\phi+B)^n\right), \tag{15}$$

$$\phi(t)=\sqrt{\frac{8}{A}}\frac{\left[A(n-1)(t+C)\right]^{(2-n)/2(1-n)}}{(2-n)}-B, \tag{16}$$

$$H(t)=\left[A(n-1)(t+C)\right]^{1/(1-n)}, \tag{17}$$

$$a(t)=a_0\exp\left\{\frac{(1-n)A\left[(n-1)\right]^{1/(1-n)}(t+C)^{(2-n)/(1-n)}}{2-n}\right\}, \tag{18}$$

$$w(t)=-1+A\left[A(n-1)(t+C)\right]^{n/(1-n)}, \tag{19}$$

where A, B, β and C are arbitrary constants, the constant n accepts the integer values ($n\neq 1,2$).

The solutions (15) – (18) correspond to the solutions received in the paper [8] with another state parameter.

The stage of inflation is provided by approximate linear dependence of kinetic energy on the state parameter and an exit from inflation is provided by means of the special choice of parameters A, B, C and n.

The state parameter as the function of time

Now we consider solutions for the state parameter as the function of time

$$w(t)=-1+\frac{A_1}{\beta(A_1t+B_1)^2}, \tag{20}$$

$$V(\phi)=(3-A_1)B_1^2\exp\left(\pm\sqrt{2A_1}\phi\right), \tag{21}$$

$$\phi(t)=\pm\sqrt{\frac{2}{A_1}}\ln\left(\frac{1}{C_1(A_1t+B_1)}\right), \tag{22}$$

$$H(t)=\frac{1}{(A_1t+B_1)}, \tag{23}$$

$$a(t)=a_0\left[(A_1t+B_1)\right]^{1/A}. \tag{24}$$

This is the model of power-law inflation which is considered in [6,14]. It is necessary the special choice of parameters for exit on the radiation domination stage, also.

The model with graceful exit

We will consider, in the context of kinetic approximation, model of cosmological inflation with graceful exit which we will set, determining the state parameter as function of time

$$w(t)=-1-\frac{2\alpha^6\lambda^2}{\beta}\coth^2\left(\frac{2\lambda t}{\sqrt{\beta}}\right)+\frac{2\alpha^2\lambda^2}{\beta}. \tag{25}$$

The solutions of equations (12) – (14) are

$$V(\phi)=\alpha^2\lambda^2\left[(3\alpha^2-2)\cosh^2\left(\frac{\phi}{\alpha}\right)+2\right], \tag{26}$$

$$\phi(t)=\alpha\ln\left[\tanh\left(\frac{\lambda t}{\sqrt{\beta}}\right)\right], \tag{27}$$

$$H(t)=\alpha^2\lambda\coth\left(\frac{2\lambda t}{\sqrt{\beta}}\right), \tag{28}$$

Journal of Physics: Conference Series **731** (2016) 012004 doi:10.1088/1742-6596/731/1/012004

$$a(t) = a_0 \sinh\left(\frac{2\lambda t}{\sqrt{\beta}}\right)^{\alpha^2/2}. \tag{29}$$

Graceful exit from inflation, reheating and further correct transition to the stage of radiation domination $(w = 1/3)$ occurs at the choice of parameter $\lambda = \pm\frac{1}{3\alpha}\sqrt{\frac{6\beta}{1-\alpha^4}}$, where $\alpha^4 < 1$.

The cosmological parameters

During inflation, quantum fluctuations of the scalar field created perturbations of metric. In a linear order, for scalar and tensor perturbations the basic parameters of cosmological perturbations on crossing of Hubble radius $(k = aH)$ for exact solutions were calculated in works [18-19]

The power spectra of scalar and tensor perturbations

$$\mathcal{P}_S(k) = -\left.\frac{H^4}{8\pi^2 \dot{H}}\right|_{k=aH}, \tag{30}$$

$$\mathcal{P}_T(k) = -\left.\frac{H^4}{2\pi^2}\right|_{k=aH}. \tag{31}$$

The spectral indexes

$$n_S(k) - 1 = \left.\frac{4\dot{H} - \frac{H\ddot{H}}{\dot{H}^2}}{\dot{H} + H^2}\right|_{k=aH}, \tag{32}$$

$$n_T(k) = \left.\frac{2\dot{H}}{\dot{H} + H^2}\right|_{k=aH}. \tag{33}$$

The tensor-to-scalar ratio

$$r = -4\frac{\dot{H}}{H^2}. \tag{34}$$

Values of cosmological parameters are limited to observational data [4]

$$10^9 \mathcal{P}_S = 2.142 \pm 0.049,$$

$$\mathcal{P}_T = r\mathcal{P}_S,$$

$$n_S = 0.9667 \pm 0.0040,$$

$$r < 0.113.$$

One can calculate parameters of cosmological perturbations from the received solutions. Also important parameter is the number of e-folds

$$N = \int_{t_i}^{t_e} H(t)dt \geq 60, \tag{35}$$

Where t_i - the time of the beginning and t_e - the time of the end of the inflationary stage.

For recalculation of the power spectra on a modern stage it is necessary to multiply their values on crossing of Hubble radius on the transfer function square. The transfer function $T(k)$ is found to be well approximated by the form [20]

Journal of Physics: Conference Series **731** (2016) 012004 doi:10.1088/1742-6596/731/1/012004

$$T(k)=\frac{3j_1(k\eta_0)}{k\eta_0}\sqrt{1+1.34\left(\frac{k}{k_{eq}}\right)+2.5\left(\frac{k}{k_{eq}}\right)^2}\,, \tag{36}$$

where η_0 - the conformal time corresponding to the modern stage, $k_{eq}=0.073\Omega_m h_0^2\ Mpc^{-1}$ is the wave number corresponding to the Hubble radius at the time of the matter-radiation equality.

Thus, the model parameters must be set taking into account the observational data.

Conclusion

In this work the method of the solution of the scalar field dynamical equations on the inflationary stage based on representation of kinetic energy by linear function of the state parameter was considered. Such representation of kinetic energy is entered for providing a stage of the accelerated expansion. Within the offered approach the model of cosmological inflation for the set state parameter with the graceful exit to a stage of radiation domination has been constructed.

Advantage of kinetic approximation approach concerning of slow roll approximation is the accounting of a scalar field kinetic energy and existence of an inflationary stage for arbitrary potentials.

Also solutions of the equations (1) – (3) for arbitrary potentials can be received by methods of exact solutions [5-17]. The existence of the stage of accelerated expansion is provided by means of kinetic approximation.

This work was supported by RFPF grants № 16-02-00488 A and № 16-08-00618 A.

References

[1] Starobinsky A A *Phys. Lett.* B **91** p 99
[2] Mukhanov V F, Feldman H A and Brandenberger R H *Phys. Rep.* **215** p 203
[3] Liddle A R and Lyth D H *Phys. Rep.* **231** p 1
[4] *Ade P et al. Planck 2015 results. XIII. Cosmological parameters, Astronomy & Astrophysics manuscript No. planck parameters* (Preprint astro-ph/1502.01589)
[5] Barrow J D *Phys. Lett.* B **187** p 12
[6] Lucchin F and Matarrese S *Phys. Rev.* **32** p 1316
[7] Barrow J D *Phys. Rev.* **49** 3055
[8] Schunck F E and Mielke E W *Phys. Rev.* **50** 4794
[9] Zhuravlev V M, Chervon S V and Shchigolev V K *JETP* **87** p 223
[10] Norbury J W *Phys. Lett.* B **433** p 263
[11] Starkovich S P and Cooperstock F I *Astrophysical Journal* **398** p 1
[12] Lidsey J E *Phys. Lett.* B **273** p 42
[13] Ellis G and Madsen M S *Class. Quantum Grav.* **8** p 667
[14] Salopek D S and Bond J R *Phys. Rev.* **42** 3936
[15] Muslimov A G *Class. Quantum Grav.* **7** p 231
[16] Lobo S N, Mak M K *Int. J. Mod. Phys.* **23** 1450063
[17] Hawkins R M, Lidsey J E *Phys.Rev.* **66** 023523
[18] Chervon S V, Novello M and Triay R *Grav. & Cosm.* **11** no. 4 (44) p 329
[19] Chervon S V, Fomin I V *Grav. & Cosm.* **14** no. 2 p 163
[20] Kosowsky A and Turner M S *Phys. Rev.* **52** 1739

Universal non-destructive testing method in the microwave range

R Yu Gerasimov, G N Fadeev, Yu V Gerasimov, E A Kondrakova
Department of Physics, Bauman Moscow State Technical University, Moscow, Russia
E-mail: drgerasimov.y@gmail.com

Abstract. Considered in this paper, a new method of nondestructive control based on the spectra of radio wave radiation of the microwave range wideband scanning receiver. It is experimentally shown that this method has a high accuracy in the determination of the frequency spectrum. This allows the method considered to reach a sensitivity of 5-6 Hz/nm.

We propose a method of non-destructive monitoring parameters (chemical, physical and mechanical) in the process of obtaining and using structural composite materials and products based on complex structured materials. A number of new tasks can be solved with the proposed method of non-destructive testing, which allows to determine the control sensitivity 5-6 Hz/nm.

This non-destructive testing method is based on recording spectra of radio-wave radiation from the test object broadband scanning receiver and subsequent analysis of the resulting spectra on PC. The stated technical problem is achieved in that the test object excite the electromagnetic field of the external electromagnetic radiation, as a result, arise macro-currents and characteristic of the capacitive element, the bias currents.

The proposed method of non-destructive testing is based [1] to the following effect. Induced electromagnetic field (Fig. 1) is excited in control object *1* by external electromagnetic radiation source *3*. This gives rise to free currents and displacement currents, which are characteristic of a capacitive element. Irradiation is performed by two flexible conductive plates through two dielectric plates *2* (acting as a sort of a capacitor), between which control object *1* is placed. The dielectric plates are laid along the outline of the control object. Because of this, the field (emitted by the object) and the spectral characteristics being determined become object oriented, the object itself becomes an electromagnetic field radiator. To position details of objects under investigation, the dielectric plates are made as a set of individual parts. To scan in homogeneities and select signals of individual elements, scanning receiver *4* is used, which is controlled by PC *5*.

The presence of two dielectric plates (through which external electromagnetic radiation source *3* excites an electromagnetic field in the control object) minimizes reflections from their external surfaces and, hence, excitation of surface waves. This improves the resolving power of the equipment in case of detecting deviations from the characteristics of a reference sample.

The main difference of this method is that the object of control acts as an antenna radiator.

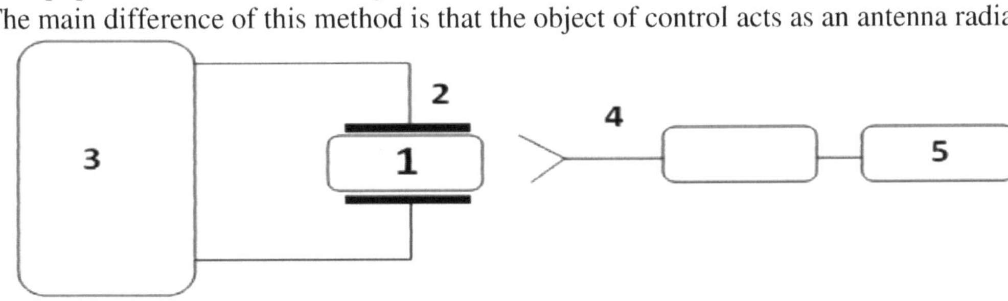

Fig. 1. Schematic of a setup: (*1*) control object, (*2*) plates of capacitive operation sensor, (*3*) high frequency voltage generator, (*4*) scanning receiver, and (*5*) PC

Let us briefly consider the physics of processes in the control object. When the generator is switched on, a high frequency current enters a central coaxial cable and passes to a lead wire connected to the object and the plates of sensor *2*. This results in the formation of standing waves.

In the dielectric media (the dielectric layer of the sensor), displacement currents appear, which rotate dipoles in the dielectric until they are aligned with the current direction. Further, the displacement current takes place in the object of control here, involving mobile electrons induced electric charge. In the center of the object, there is a charge of the same sign, and the edges of the bias current has the opposite sign. Thus, at the edges and the center of the liquid, charges of the opposite signs accumulate, because of which a T-type wave forms. The T-type waves travelling from the center to the edges and the standing waves produced at the lead wires generate a varying electromagnetic field around the object. The spectrum of this field characterizes the control object. By comparing the spectrum of the control object with the spectrum of the reference sample, the changed parameters can be estimated.

As a result of experiments were obtained spectra of defective composite panels . Such defects can be: starved spot, additional inclusion, local destruction and etc. On the spectra one can observe clear differences in the spectral pattern and to detect the presence of defects, which were not included in the design.

As a result the processing of the envelopes of the spectra were obtained (Fig. 2) graphics for defects of various sizes.

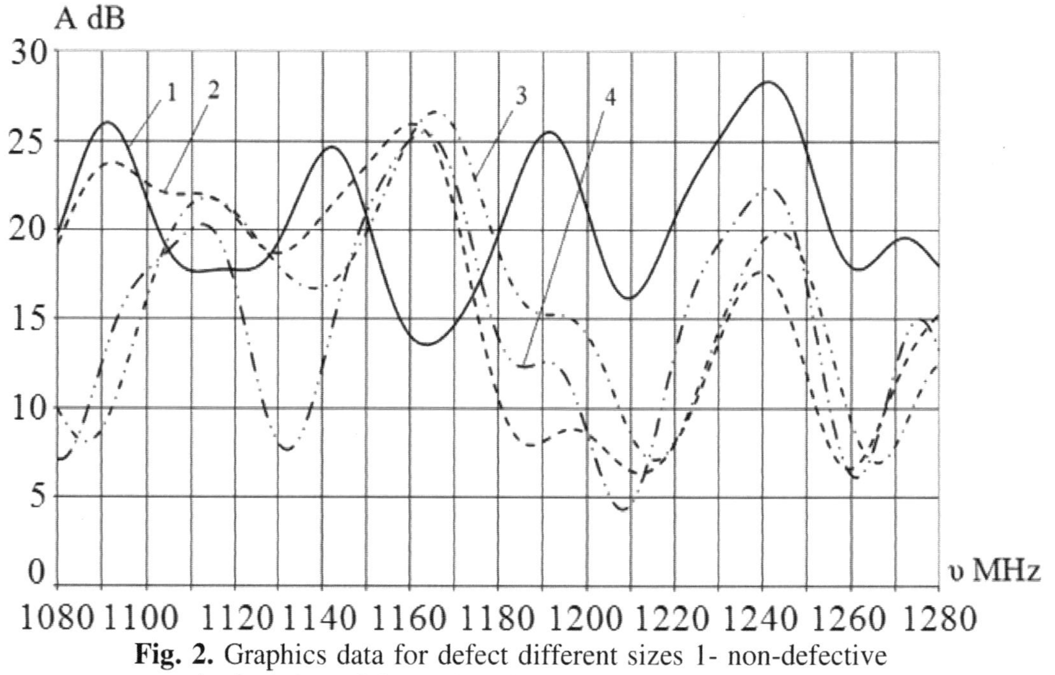

Fig. 2. Graphics data for defect different sizes 1- non-defective sample; 2- defect of 6 mm;3- defect is 8 mm; 4- defect of 10 mm

As a result of comparison you can determine the size and location of the defect. You can see that the spectrum reflecting the highs offset the defects on the period in relation to the chart reflecting the non-defective sample. Charts differ in amplitude, which also allows to distinguish the size and position relative to the capacitive sensors of the defect.

For the determine the accuracy of the method of non-destructive testing were conducted experiments on the powders [2,3]. During the experiments were obtained the spectra of powders of boron carbide, titanium dioxide, silicon carbide. The size of the particles ranged from 3 to 63 μm. On Fig. 3 shown the spectra of electromagnetic radiation at continuous irradiation of samples of micro-powders of silicon carbide with different predominant particle size from 3-5 up to 50-63 microns when excited in the capacitive resonator at the frequency ν=2.4 GHz.

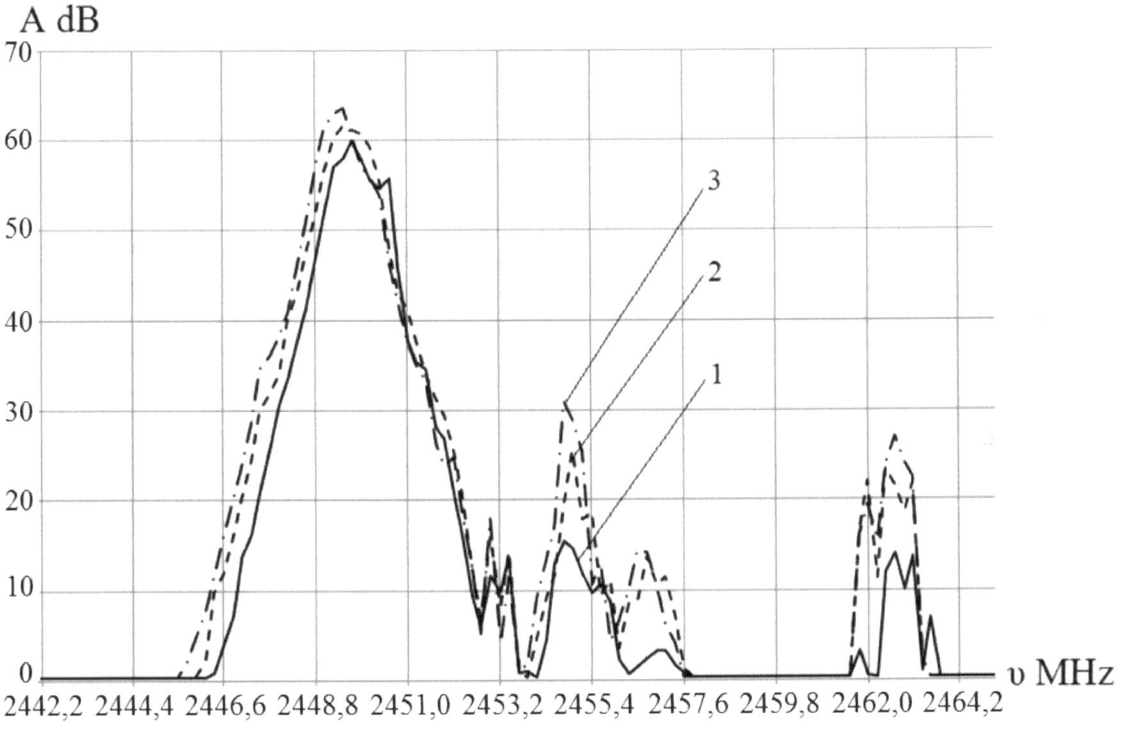

Fig. 3. The spectral characteristics of silicon carbide powders having a particle size of 3-5 μm (1), 20-28 μm (2), 50-63 μm(3)

In all cases, when entering into the resonator of the sample powders were modified forms of the spectra. The General trend is: as you increase the size of powder particles was the reduction of the peak frequency and increasing amplitude, which was reflected in the band offsets of the spectrum. On the graphs of the spectra and the sizes of powder particles has turned out to determine the sensitivity equal to (5-6).109Гц/m = 5-6 Hz/nm at the resolution of the receiver by a frequency of Δν=0,01-0,10 Hz.

References

[1] Gerasimov R Yu , Fadeev G N, Gerasimov Yu V, Kondrakova E A 2015 Identification of deuterium and tritium isotopes in water by microwave radioscopy *Doklady Physical Chemistry* **464** pp 219–221
[2] Bovenko V N and Gerasimov R Yu 2011 Formation of spectral characteristics of signals on irradiation of micropowders in the microwave range *Nonliner world* **6** pp 369–377
[3] Bovenko V N, Gerasimov R Yu and Gorobets, L Zh 2010 Radio spectroscopy of Finely Ground Products *Scientific Session of NIYaU MIFI_2010 Nanophysics and Nanotechnologies, Basic Problems of Science* **2** pp 73–74

Equation of state for hydrogen-like materials with high energy densities

Y V Gerasimov[1] and A G Maslov[2]
[1] Associated professor, Bauman Moscow State Technical University, Moscow, Russia
[2] Assistant professor, Bauman Moscow State Technical University, Moscow, Russia
E-mail: drgerasimov.y@gmail.com, maslov_art@bmstu.ru

Abstract. Solution of time-dependent Schrödinger equation for deuterium with high density and pulse impact on a system is described in the article. Result compared with experiments of shock compression of deuterium, made by RFNC-VNIIEF, Sandia and NOVA. The way of use of this solution in advanced railgun technologies is considered.

1. Introduction

The problem of pulse thermonuclear fusion isn't finally solved yet. It is problem with high energy densities, high pressures and temperatures. It is necessary to know equations of state for different materials in this extreme impact. Parameters are about 1..5 TPa for pressure, and billions degrees of temperature. For processes of nanosatellite acceleration (movement of the plasma piston in the railgun and acceleration by it a nanosatellite) these parameters are some lower, but still higher than then parameters in existing experimental and theoretical data. On the other hand, by the reasons mentioned below this problem have to be solved from the quantum point of view.

2. Solving of the time-dependent Schrödinger equation for atoms of impacted material

The central problem of the article is the behavior of the atoms in the material under impact. It is interesting to consider impacts described in [1-9]. De Broglie wavelength of material's atoms in such impact has to be about 20 pm and the interplanar spacing in deuterium crystal is about 220 pm; that's why the time-dependent Schrödinger equation is used. Material is solid in the beginning of the process, so the classic potential of solid-state physics is used. However, matter wavelength of electrons should be 72 nm, so they are not included in the system; hence the Schrödinger equation for this system will be:

$$i\hbar\frac{\partial\varphi(\vec{r}_1,...,\vec{r}_N,t)}{\partial t}+\frac{\hbar^2}{2m}\sum_{k=1}^{N}\nabla^2{}_k\varphi(\vec{r}_1,...,\vec{r}_N,t)-\frac{e^2}{2\varepsilon_0}\sum_{k=1}^{N}\sum_{\substack{j=1\\ j\neq k}}^{N}\frac{1}{r_{kj}}\varphi(\vec{r}_1,...,\vec{r}_N,t)= \\ =V(\vec{r}_1,...,\vec{r}_N,t)\varphi(\vec{r}_1,...,\vec{r}_N,t), \tag{1}$$

where e is the electron charge, ∇^2 – Laplace operator, φ – wave function, r_{ij} - distance between 2 particles, m – mass of material's atom, i – complex unit, ε_0 – vacuum permittivity.

Approximation of exchange energy is used:

$$\varphi(\vec{r}_1,...,\vec{r}_N,t)=\prod_{k=1}^{N}\psi(\vec{r}_k,t),$$

after differentiation it by time in the equation (1)

Journal of Physics: Conference Series **731** (2016) 012006 doi:10.1088/1742-6596/731/1/012006

$$i\hbar\sum_{l=1}^{N}\prod_{\substack{j=1\\ j\neq i}}^{N}\psi(\vec{r}_j,t)\frac{\partial}{\partial t}\psi(\vec{r}_l,t)+\frac{\hbar^2}{2m}\sum_{l=1}^{N}\prod_{\substack{j=1\\ j\neq i}}^{N}\psi(\vec{r}_j,t)\nabla^2{}_l\psi(\vec{r}_l,t)-\frac{e^2}{2\varepsilon_0}\sum_{l=1}^{N}\sum_{\substack{j=1\\ j\neq i}}^{N}\prod_{\substack{k=1\\ k\neq i}}^{N}\psi(\vec{r}_k,t)\frac{1}{r_{lj}}\psi(\vec{r}_l,t)=$$
$$=\prod_{\substack{j=1\\ j\neq l}}^{N}\psi(\vec{r}_j,t)V(\vec{r}_1,...,\vec{r}_N,t)\psi(\vec{r}_l,t) \quad (2)$$

to simplify equation was postulated vectors

$$\vec{\Phi}(\vec{r}_1,...,\vec{r}_N,t)=\begin{pmatrix}\psi(\vec{r}_1,t)\\ \vdots\\ \psi(\vec{r}_N,t)\end{pmatrix},\ \vec{\phi}=\begin{pmatrix}\prod_{k=2}^{N}\psi(\vec{r}_k,t), & \dots & \prod_{\substack{k=1\\ k\neq j}}^{N}\psi(\vec{r}_k,t), & \dots & \prod_{k=1}^{N-1}\psi(\vec{r}_k,t)\end{pmatrix},$$

and matrix operators

$$\hat{A}=\begin{pmatrix}\nabla^2{}_1 & 0 & \cdots & 0\\ 0 & \ddots & \ddots & \vdots\\ \vdots & \ddots & \ddots & 0\\ 0 & \cdots & 0 & \nabla^2{}_N\end{pmatrix}_{N\times N},\ \hat{R}=\begin{pmatrix}0 & \frac{1}{r_{12}} & \cdots & \frac{1}{r_{1N}}\\ \frac{1}{r_{12}} & 0 & \ddots & \vdots\\ \vdots & \ddots & 0 & \frac{1}{r_{N-1N}}\\ \frac{1}{r_{1N}} & \cdots & \frac{1}{r_{N-1N}} & 0\end{pmatrix}_{N\times N}.$$

In this case, equation (2) has the following view:

$$i\hbar\vec{\phi}\frac{\partial\vec{\Phi}(\vec{r}_1,...,\vec{r}_N,t)}{\partial t}+\frac{\hbar^2}{2m}\vec{\phi}\hat{A}\vec{\Phi}(\vec{r}_1,...,\vec{r}_N,t)-\frac{e^2}{2\varepsilon_0}\vec{\phi}\hat{R}\vec{\Phi}(\vec{r}_1,...,\vec{r}_N,t)=\vec{\phi}V(\vec{r}_1,...,\vec{r}_N,t)\vec{\Phi}(\vec{r}_1,...,\vec{r}_N,t).(3)$$

It is useful to find the fundamental solution for this equation. It means that V should be unit matrix in the 0 point of space and time, so

$$\frac{\partial\Phi(\vec{r}_1,...,\vec{r}_N,t)}{\partial t}-\frac{i\hbar}{2m}\hat{A}\Phi(\vec{r}_1,...,\vec{r}_N,t)+\frac{ie^2}{2\hbar\varepsilon_0}\hat{R}\Phi(\vec{r}_1,...,\vec{r}_N,t)=I\delta(\vec{r}_1,...,\vec{r}_N,t)\Phi(\vec{r}_1,...,\vec{r}_N,t).\ (4)$$

Where $\Phi(\vec{r}_1,...,\vec{r}_N,t)$ is the fundamental solution.

In wave vector space, where $\vec{\eta}_i$ – wave vectors, Fourier transfer – $\hat{F}$

$$\hat{F}\left(\frac{1}{x}\right)=i\pi sign(\eta_1);$$

$$\hat{F}\left(\frac{1}{|\vec{r}|}\right)=\left|\begin{pmatrix}i\pi sign(\eta_1)\\ i\pi sign(\eta_2)\\ i\pi sign(\eta_3)\end{pmatrix}\right|=i\pi\sqrt{\sum_i\left(sign(\eta_i)\right)^2}=i\pi\sqrt{3};$$

$$\hat{F}\left(\hat{R}\right)=\begin{pmatrix} 0 & i\pi\sqrt{3} & \ldots \\ i\pi\sqrt{3} & 0 & i\pi\sqrt{3} \\ \ldots & i\pi\sqrt{3} & 0 \end{pmatrix}_{N\times N};$$

$$\hat{F}\left(\hat{A}\right)=\begin{pmatrix} \eta_1^{\,2} & 0 & 0 & 0 & 0 \\ 0 & \ldots & 0 & 0 & 0 \\ 0 & 0 & \eta_i^{\,2} & 0 & 0 \\ 0 & 0 & 0 & \ldots & 0 \\ 0 & 0 & 0 & 0 & \eta_N^{\,2} \end{pmatrix}_{N\times N},$$

so the equation will be

$$\frac{\partial\tilde{\Phi}(\vec{\eta},t)}{\partial t}-\frac{i\hbar}{2m}\begin{pmatrix} \vec{\eta}_1^{\,2} & 0 & 0 & 0 & 0 \\ 0 & \ldots & 0 & 0 & 0 \\ 0 & 0 & \vec{\eta}_i^{\,2} & 0 & 0 \\ 0 & 0 & 0 & \ldots & 0 \\ 0 & 0 & 0 & 0 & \vec{\eta}_N^{\,2} \end{pmatrix}_{N\times N}\tilde{\Phi}(\vec{\eta}_1,\ldots,\vec{\eta}_N,t)-\frac{e^2\pi\sqrt{3}}{2\hbar\varepsilon_0}\begin{pmatrix} 0 & 1 & \ldots \\ 1 & 0 & 1 \\ \ldots & 1 & 0 \end{pmatrix}_{N\times N}\tilde{\Phi}(\vec{\eta}_1,\ldots,\vec{\eta}_N,t)= \quad (5)$$
$$=I\delta(t)\tilde{\Phi}(\vec{\eta}_1,\ldots,\vec{\eta}_N,t).$$

Finally in wave vectors space

$$\frac{\partial\tilde{\Phi}(\vec{\eta}_1,\ldots,\vec{\eta}_N,t)}{\partial t}-\begin{pmatrix} \frac{i\hbar}{2m}\vec{\eta}_1^{\,2} & 0 & 0 & 0 & 0 \\ 0 & \ldots & 0 & 0 & 0 \\ 0 & 0 & \frac{i\hbar}{2m}\vec{\eta}_i^{\,2} & 0 & 0 \\ 0 & 0 & 0 & \ldots & 0 \\ 0 & 0 & 0 & 0 & \frac{i\hbar}{2m}\vec{\eta}_N^{\,2} \end{pmatrix}_{N\times N}\tilde{\Phi}(\vec{\eta}_1,\ldots,\vec{\eta}_N,t)-$$
$$-\begin{pmatrix} 0 & \frac{e^2\pi\sqrt{3}}{2\hbar\varepsilon_0} & \ldots \\ \frac{e^2\pi\sqrt{3}}{2\hbar\varepsilon_0} & 0 & \frac{e^2\pi\sqrt{3}}{2\hbar\varepsilon_0} \\ \ldots & \frac{e^2\pi\sqrt{3}}{2\hbar\varepsilon_0} & 0 \end{pmatrix}_{N\times N}\tilde{\Phi}(\vec{\eta}_1,\ldots,\vec{\eta}_N,t)=I\delta(t)\tilde{\Phi}(\vec{\eta}_1,\ldots,\vec{\eta}_N,t).$$

If the following matrix is designated as B

$$B=\begin{pmatrix} \frac{i\hbar}{2m}\vec{\eta}_1^{\,2} & \frac{e^2\pi\sqrt{3}}{2\hbar\varepsilon_0} & \cdots & \cdots & \frac{e^2\pi\sqrt{3}}{2\hbar\varepsilon_0} \\ \frac{e^2\pi\sqrt{3}}{2\hbar\varepsilon_0} & \cdots & \frac{e^2\pi\sqrt{3}}{2\hbar\varepsilon_0} & \cdots & \cdots \\ \cdots & \frac{e^2\pi\sqrt{3}}{2\hbar\varepsilon_0} & \frac{i\hbar}{2m}\vec{\eta}_i^{\,2} & \frac{e^2\pi\sqrt{3}}{2\hbar\varepsilon_0} & \cdots \\ \cdots & \cdots & \frac{e^2\pi\sqrt{3}}{2\hbar\varepsilon_0} & \cdots & \frac{e^2\pi\sqrt{3}}{2\hbar\varepsilon_0} \\ \frac{e^2\pi\sqrt{3}}{2\hbar\varepsilon_0} & \cdots & \cdots & \frac{e^2\pi\sqrt{3}}{2\hbar\varepsilon_0} & \frac{i\hbar}{2m}\vec{\eta}_N^{\,2} \end{pmatrix}_{N\times N}$$

The equation (5) looks like that:

$$\frac{\partial\tilde{\Phi}(\vec{\eta}_1,...,\vec{\eta}_N,t)}{\partial t}-B\tilde{\Phi}(\vec{\eta}_1,...,\vec{\eta}_N,t)=I\delta(t)\tilde{\Phi}(\vec{\eta}_1,...,\vec{\eta}_N,t)\,. \tag{6}$$

The initial conditions for fundamental equation are unit

$$\tilde{\Phi}\left(\vec{\eta}_1,...,\vec{\eta}_N,0\right)=\frac{1}{\sqrt{N}}\begin{pmatrix}1\\ \vdots\\ 1\end{pmatrix}=\vec{D}\,,$$

and the solution of (6) is

$$\tilde{\Phi}=C\cdot e^{Bt}$$

with initial conditions

$$\tilde{\Phi}=\vec{D}\cdot e^{Bt}\,.$$

To introduce the solution in wave vectors explicitly, the definition of matrix function is used. Thus, it is necessary to calculate eigenvalues of B. The omission of the eigenvalues calculation can be written as:

$$\lambda_i=\frac{i\hbar}{2m}\vec{\eta}_i^{\,2}+\frac{e^2\pi\sqrt{3}}{2\hbar\varepsilon_0}\Theta\left(\frac{e^2\pi\sqrt{3}}{2\hbar\varepsilon_0}-\frac{i\hbar}{2m}\vec{\eta}_i^{\,2}\right).$$

Therefore, the fundamental solution is the inverse Fourier transform:

$$\Phi(\vec{r}_1,...,\vec{r}_N,t)=F^{-1}\left(\tilde{\Phi}(\vec{\eta}_1,...,\vec{\eta}_N,t)\right)=\frac{1}{(2\pi)^{3N}}F\left(\tilde{\Phi}(\vec{\eta}_1,...,\vec{\eta}_N,t)\right)=\frac{1}{(2\pi)^{3N}}\vec{D}\int_{V_\eta}e^{\mathrm{B}t}e^{i\vec{r}_j\vec{\eta}_j}dV_\eta=$$

$$=\frac{1}{(2\pi)^{3N}}\vec{D}\int_{V_\eta}T(\vec{\eta}_1,...,\vec{\eta}_N)\begin{pmatrix} e^{\lambda_1 t+i\vec{r}_1\vec{\eta}_1} & 0 & ... & ... & 0 \\ 0 & ... & 0 & ... & ... \\ ... & 0 & e^{\lambda_j t+i\vec{r}_j\vec{\eta}_j} & 0 & ... \\ ... & ... & 0 & ... & 0 \\ 0 & ... & ... & 0 & e^{\lambda_N t+i\vec{r}_N\vec{\eta}_N} \end{pmatrix}T^T(\vec{\eta}_1,...,\vec{\eta}_N)dV_\eta= \quad (7)$$

$$=\frac{1}{(2\pi)^{3N}}e^{\frac{e^2\pi\sqrt{3}}{2h}t}\vec{D}\int_{V_\eta}T(\vec{\eta}_1,...,\vec{\eta}_N)\begin{pmatrix} e^{\frac{ih}{2m}\vec{\eta}_j^{\,2}t+i\vec{r}_j\vec{\eta}_j} & 0 & 0 \\ 0 & \ddots & 0 \\ 0 & 0 & e^{\frac{ih}{2m}\vec{\eta}_j^{\,2}t+i\vec{r}_j\vec{\eta}_j} \end{pmatrix}T^T(\vec{\eta}_1,...,\vec{\eta}_N)dV_\eta.$$

So the fundamental solution of equation (4) is the superposition of N solutions, which describes N plane waves; it is analogical to Bloch waves. In these waves the dispersion relations are:

$$\omega_j=\frac{\hbar}{2m}\vec{\eta}_j^{\,2}. \quad (8)$$

Phase of the wave depends on wave vector and coordinates:

$$\gamma_j=\vec{r}_j\vec{\eta}_j+\frac{\hbar}{2m}\vec{\eta}_j^{\,2}t. \quad (9)$$

Phase and group speeds of the waves:

$$c_i^{Ph}=\frac{\hbar}{2m}|\vec{\eta}_i|, \quad (10)$$

$$c_i^{Gr}=\frac{\hbar}{2m}2|\vec{\eta}_i|=\frac{\hbar}{m}|\vec{\eta}_i|. \quad (11)$$

To express the solution of fundamental equation explicitly, we need to use Radon transform. On fixed direction of wave packet solution is integrated by the wave vector scalar. It is useful to designate as α, β_j and $\vec{æ}_j$ the following designations:

$$\alpha=\frac{i\hbar}{2m}t,\ \beta_j=i\vec{r}_j\vec{æ}_j,\ \vec{æ}_j=\frac{\overrightarrow{\eta_j}}{\left|\overrightarrow{\eta_j}\right|}$$

And integrate in spatial angle

$$\Phi(\vec{r}_1,...,\vec{r}_N,t)=\frac{1}{(2\pi)^{3N}}\dot{D}e^{\frac{e^2\pi\sqrt{3}}{2h}t}\cdot$$

$$\cdot\int_{\Omega}T(\vec{æ}_1,...,\vec{æ}_N)\begin{pmatrix}\int_0^\infty e^{\alpha|\vec{\eta}_1|^2+\beta_1|\vec{\eta}_1|}d|\vec{\eta}_1| & 0 & ... & ... & 0\\ 0 & \ddots & 0 & ... & ...\\ ... & 0 & e^{\alpha|\vec{\eta}_j|^2+\beta_j|\vec{\eta}_j|}d|\vec{\eta}_j| & 0 & ...\\ ... & ... & 0 & \ddots & 0\\ 0 & ... & ... & 0 & e^{\alpha|\vec{\eta}_N|^2+\beta_N|\vec{\eta}_N|}d|\vec{\eta}_N|\end{pmatrix}T^T(\vec{æ}_1,...,\vec{æ}_N)d\Omega. \quad (12)$$

Let us consider the integral by wave vector:

$$\int_0^\infty e^{\alpha|\vec{\eta}_j|^2+\beta_j|\vec{\eta}_j|}d|\vec{\eta}_j|=\left\||\vec{\eta}_j|=z\right|=\int_0^\infty e^{\alpha z^2+\beta_j z}dz=\int_0^\infty e^{\alpha\left(z^2+\frac{\beta_j}{\alpha}z\right)}dz=$$
$$=\int_0^\infty e^{\alpha\left(z^2+\frac{\beta_j}{\alpha}z+\left(\frac{\beta_j}{2\alpha}\right)^2-\left(\frac{\beta_j}{2\alpha}\right)^2\right)}dz=\int_0^\infty e^{\alpha\left(z+\frac{\beta_j}{2\alpha}\right)^2-\frac{\beta_j^2}{4\alpha}}dz=e^{-\frac{\beta_j^2}{4\alpha}}\int_0^\infty e^{\alpha\left(z+\frac{\beta_j}{2\alpha}\right)^2}dz=A_je^{-\frac{im(\vec{r}_j\vec{æ}_j)^2}{2\hbar t}}. \quad (13)$$

To calculate A_j we have to designate parameters once again:

$$a=\frac{\hbar}{2m}t,\ z_0=\frac{2m\vec{r}_j\vec{æ}_j}{\hbar t},$$

therefore

$$A_j=\int_0^\infty e^{ia(z+z_0)^2}dz=\lim_{z\to\infty}\frac{1}{2}\frac{\sqrt{i\pi}\left(\operatorname{erf}\left(\sqrt{\frac{a}{i}}(z+z_0)\right)-\operatorname{erf}\left(\sqrt{\frac{a}{i}}z_0\right)\right)}{\sqrt{a}}. \quad (14)$$

It is possible to show that $\lim_{z\to\infty}\left(\operatorname{erf}\left(\sqrt{\frac{a}{i}}(z+z_0)\right)\right)=1$. In this case (14) transforms in:

$$A_j(\vec{r}_j\vec{æ}_j,t)=\frac{\sqrt{i\pi}\left(1-\operatorname{erf}\left(\sqrt{\frac{a}{i}}z_0\right)\right)}{2\sqrt{a}}=\frac{\sqrt{i\pi}\left(1-\operatorname{erf}\left(\sqrt{\frac{2m}{i\hbar t}}\vec{r}_j\vec{æ}_j\right)\right)}{\sqrt{\frac{2\hbar}{m}t}}. \quad (15)$$

Thus, (12) has the following view:

$$\Phi\left(\vec{r}_1,...,\vec{r}_N,t\right)=\frac{1}{\left(2\pi\right)^{3N}}\vec{D}e^{-\frac{e^2\pi\sqrt{3}}{2\hbar}t}\cdot$$

$$\cdot\int_{\Omega}T\left(\vec{æ}_1,...,\vec{æ}_N\right)\begin{pmatrix}\mu_1\left(\vec{r}_1\vec{æ}_1,t\right) & 0 & \ldots & \ldots & 0\\ 0 & \ddots & 0 & \ldots & \ldots\\ \ldots & 0 & \mu_j\left(\vec{r}_j\vec{æ}_j,t\right) & 0 & \ldots\\ \ldots & \ldots & 0 & \ddots & 0\\ 0 & \ldots & \ldots & 0 & \mu_N\left(\vec{r}_N\vec{æ}_N,t\right)\end{pmatrix}T^T\left(\vec{æ}_1,...,\vec{æ}_N\right)d\Omega,$$

where

$$\mu_j\left(\vec{r}_j\vec{æ}_j,t\right)=e^{-\frac{im\left(\vec{r}_j\vec{æ}_j\right)^2}{2\hbar t}}\frac{\sqrt{i\pi}\left(1-\operatorname{erf}\left(\sqrt{\frac{2m}{i\hbar t}}\vec{r}_j\vec{æ}_j\right)\right)}{\sqrt{\frac{2\hbar}{m}t}}.$$

Finally

$$\Phi\left(\vec{r}_1,...,\vec{r}_N,t\right)=\frac{1}{\left(2\pi\right)^{3N}}\vec{D}e^{\frac{e^2\pi\sqrt{3}}{2\hbar}t}\int_{\Omega}\sum_j\left(\left(\vec{\xi}_j\right)^2\mu_j\left(\vec{r}_j\vec{æ}_j,t\right)\right)d\Omega. \tag{16}$$

This is the fundamental solution of Schrödinger equation in plane waves for atoms of the material under the shock impact, according to Radon transform. It is possible to solve various problems by convolution with different impacts, and solving the Cauchy problem.

3. Pressure in deuterium with high density calculations

Instead of other objectives, the problem of average pressure calculations of highly dense material are possible to resolve with fundamental solution (16), in case of correct start positions of particles, equal to material density.

Algorithm of calculation based on the assumption that only the chosen particle moves, at first step calculates new average coordinates of particle, in case of the wave vector is constant, on the second step wave vector calculates for new coordinates of particle, and the same for all particles on the time layer. Calculation of average parameters is standard for quantum mechanics. This algorithm is shown of Fig. 1.

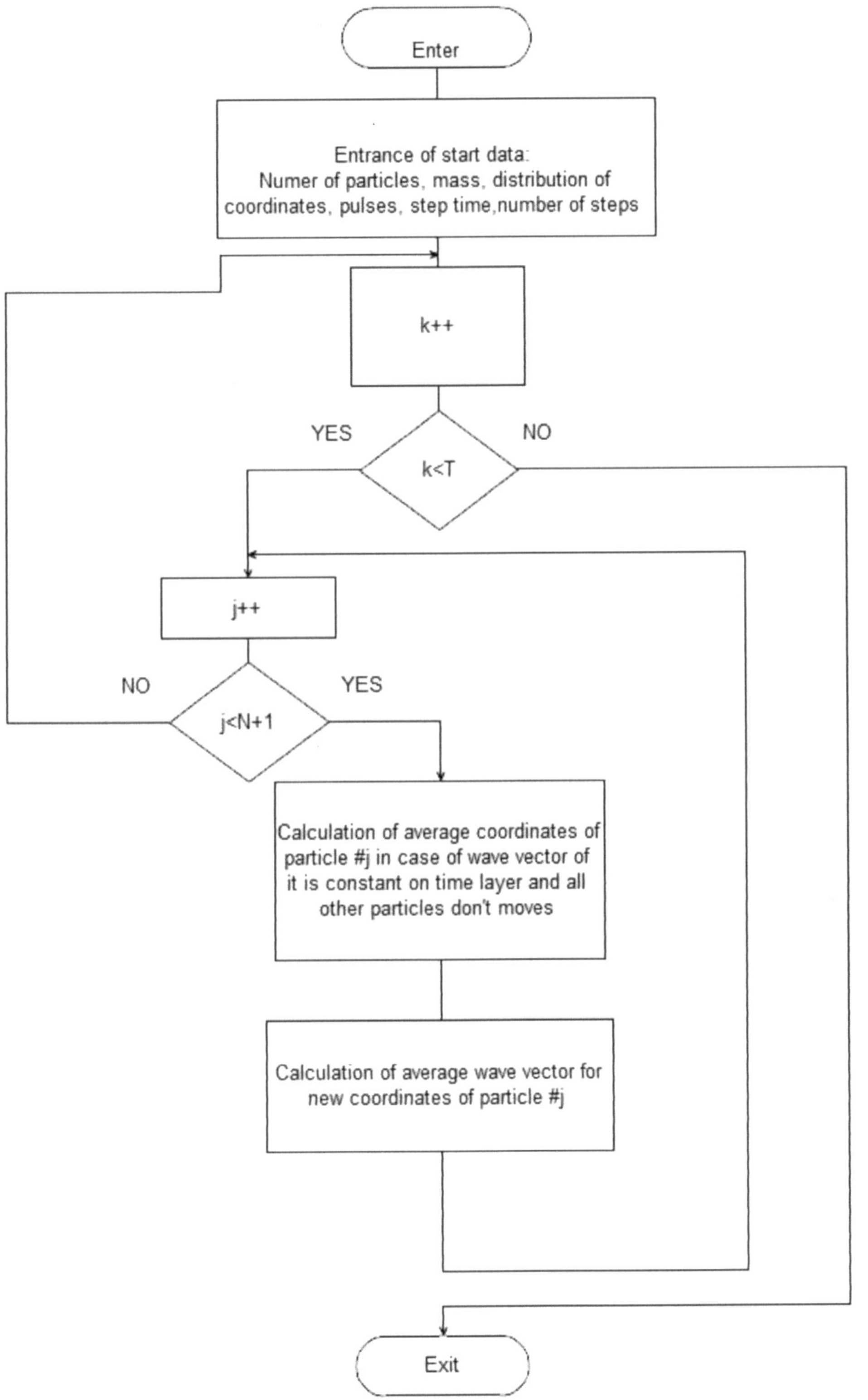

Figure 1. Principal scheme of the calculation algorithm

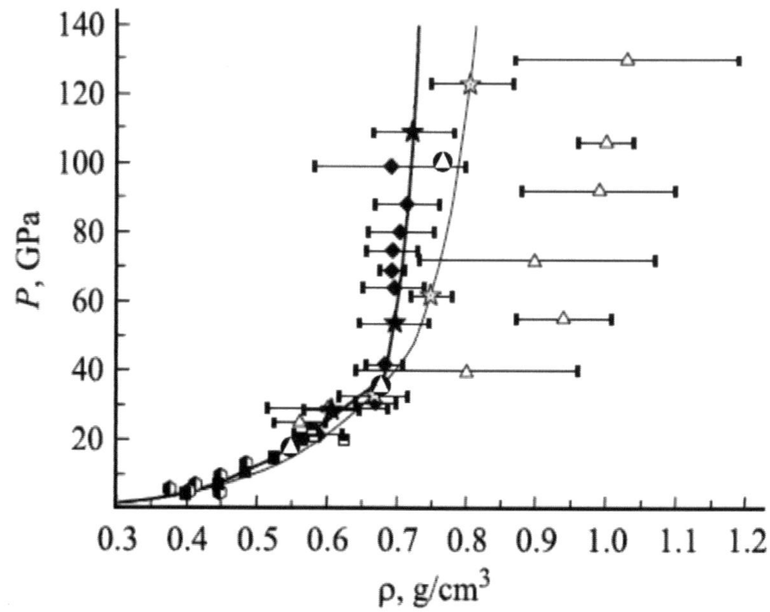

Figure 2. Experimental data and calculations for shock compression of deuterium. Data: ◐ – [1], ■ – [2], ⊟ – [3], △ – [4], ◆ – [5], ★ – [6], ☆ – [7,8,9], ⭘ – calculation in this article.

Experimental results of shock compression of deuterium by VNIIEF-RFNC, Sandia and Nova and results of the calculation are shown on Fig. 2. It is easy to see, that the results of calculation are within the experimental error.

Estimation of the number defects in crystal of nanosatellite's hull by penetration of plasma piston ions in case of a nanosatellite acceleration by railgun gives about $10^{19}..10^{21}$ defects, with energy distribution like in [10]; depth of penetration is few millimeters, that means, that in the tail of a nanosatellite have to be the number of micro-cracks, and formation of extending crack is possible while the shock unloading. It means that about 40% of energy is used on formation of defects in process of acceleration. Convolution of (16) with described impact will allow constructing material with a structure, not effected to problem, that will successfully increase the efficiency of advanced railguns.

References

[1] Dick R D, Kerley G I 1980 *J. Chem. Phys.* 73 5264
[2] Nellis W J, Mitchell A C, van Thiel L M et al. 1983 *J. Chem. Phys.* 79 1480
[3] Van Thiel L M, Ross M, Hord B L et al. 1973 *Phys. Rev. Lett.* 31 979
[4] Da Silva L B, Celliers P, Collins G W et al. 1997 *Phys. Rev. Lett.* 78 483
[5] Knudson M D, Hanson D L, Bailey J E et al. 2004 *Phys. Rev. B.* 69 144 209
[6] Boriskov G V, Bykov A I, Nellis W J et al. 2005 *Phys. Rev. B.* 71 092 104
[7] Belov S I, Boriskov G V, Bykov A I, et al. 2002 *Technical Physics. The Russian Journal of Applied Physics Lett.* 76 508
[8] Belov S I, Boriskov G V, Bykov A I et al. 2003 *Doklady Physics* 76 508
[9] Trunin R F, Boriskov G V, Bykov A I, et. al. 2006 *Technical Physics. The Russian Journal of Applied Physics* 51 7 907
[10] Gerasimov Y V, Maslov A G 2015 *Herald of the Bauman Moscow State Technical University. Series Natural Sciences* 1 69

Journal of Physics: Conference Series **731** (2016) 012007 doi:10.1088/1742-6596/731/1/012007

Axion-photon conversion in space and in low symmetrical dielectric crystals

V S Gorelik [1,2]
[1]P.N. Lebedev Physical Institute of Russian Academy of Sciences, Moscow, Russia
[2]Bauman Moscow State Technical University, Moscow, Russia
E-mail: gorelik@sci.lebedev.ru

Abstract. The opportunities of axions detection as the result of axion-photon conversion processes in the space and in low symmetrical dielectric crystals are discussed. In accordance with the modern theory predictions, axions are pseudoscalar vacuum particles having very small (0.001-1.0 meV) rest energy. The possibility of axions conversion into photons and vice-versa processes in vacuum at the presence of outer magnetic field has been analyzed before. Pseudoscalar (axion type) modes are existing in some types of crystals. Polar pseudoscalar lattice and exciton modes in low symmetrical crystals are strongly interacted with axions. In this work, optical excitation of axion-type modes in low symmetrical crystals is proposed for observation of axion – photon conversion processes. Instead of outer magnetic field, the crystalline field of such crystals may be used. The experimental schemes for axion-photon conversion processes observation with recording the secondary emission of luminescence, infrared or Stimulated Raman Scattering in some dielectric crystals are discussed.

According to the modern high energy physics theory [1-4], the Universe undergone the phase transition from homogeneous to local density space ordering state. The soft mode, responsible for symmetry lowering of vacuum, is predicted. Such mode should be total symmetrical one and should be described by scalar field with corresponding scalar elementary particles – "amplitudons" (Higgs bosons) [4-7]. Beside this, T-inversion violation in high energy physics resulted in the conclusion of the pseudoscalar elementary particles (Nambu – Goldstone bosons or "phasons" particles) with nonzero rest mass existence. Now such pseudoscalar particles are known as axions: Weakly Interacting Subelectronvolt Particles (WISP) [8-11]. WISP is expected to have the very small rest energy: 0.001...1.0 meV [12,13] The axions as pseudoscalar particles are described by wave function, changing the sign under inversion of space and mirror reflections transformations. The actual topics of modern physics is the prediction of the presence in Universe the dark matter (total share 0,23) and dark energy (total share of 0.73). The astrophysical data result in the conclusion [12-16] that one of the candidates for dark matter elemental particles are axions — pseudoscalar vacuum particles with the very low rest mass and relativistic dispersion law – WISP. "Cold" (the slowest) axions are nonrelativistic (Newtonian) particles. "Hot" (fast) axions are relativistic and move with the velocity, close to the velocity C_0 of light in vacuum. There are proposals of other candidates of dark matter elemental particles: WIMP (Weakly Interacting Massive Particles), having the rest energy about 1.0 keV; neutrino; neutralino and so on. According to the estimates obtained from astrophysical data, the equilibrium concentration of axions in our part of the galaxy is about 10^{-24}g/cm3. Accordingly axions with the lowest rest mass must be in Bose-Einstein condensate state even at room temperature. Now the opportunity of axions detection and generation in the presence of strong external magnetic field is discussed [14-18]. In this case explores two effects: 1) the detection of photons, emerging as the result of relativistic axion - photon conversion at strong magnetic field in vacuum; 2) the detection of photons, emerging as the result of Newtonian axion - photon conversion at strong magnetic field. In [17-20] on the dynamics of stars with masses 8...12 Solar masses, the conclusion was made about the processes of photon-

Journal of Physics: Conference Series **731** (2016) 012007 doi:10.1088/1742-6596/731/1/012007

axion conversion inside stars at the presence of strong magnetic field and was estimated the efficiency of photon-axion conversion for such processes. In works [14-18] it was suggested the possibility of "hot" axions creating with energy 2-3 eV in the laboratory on the base of "Light shining through wall" (Primakov-effect) experiments. The powerful laser source and strong outer magnetic field have been used. However the secondary signal of photons in "Light shining through wall" experiments was very small (~10^{-8} 1/s) for hopeful detection. In this paper we analyze the opportunity of axion-photon and photon–axion conversion processes observation in dielectric media, recently proposed in works [21-30]. The new experimental schemes to implement the generation and detection of axions in the laboratory are presented. In accordance with the known group theory, phonons and excitons states of molecules and crystals are classified by irreducible presentations (races) of the corresponding symmetry groups. Pseudo scalar presentations have characters equal to unity for proper rotations and to minus unity for improper one (inversion, mirror place and mirror rotation). We have analyzed the selection rules for one-photon processes: infrared (IR) and ultraviolet (UV) emission or absorption, and for two-and three-photon effects: Raman and Hyper Raman scattering, Two-Photon Emission and Two-Photon Excited Luminescence, for a number of liquids (water, ethanol, glycerin) and crystals (sodium nitrite, aluminum oxide, calcite, lithium niobate and some amino acid crystals). In these media pseudo scalar modes are presented. In some substances pseudo scalar modes are "silent", i.e., are forbidden as for Raman and also for one-photon (IR or UF) spectra. In several crystals the excitation or destruction of pseudo scalar state may be permitted by selection rules as a result of one photon processes (absorption or emission). Such situation takes place for C_{2h} point group (see Table 1). From this table we can see that "Z"–type polar phonon (polariton) relates to the same irreducible presentations as pseudoscalar one. Thus under resonant intersection between polariton dispersion curve ($\omega^2 = \frac{C_0^2 k^2}{\varepsilon(\omega)}$) and axion dispersion branch ($\omega^2 = \omega_a^2 + C_0^2 k^2$) the polariton-axion conversion processes should be realized without applications of outer magnetic field but due to the lowering of the symmetry as a result of crystalline field presence, corresponding to C_{2h}–point group. The examples of crystals, having C_{2h}–point group symmetry, are stilbene, antracene, KGW and others. So the excitation of A_2– type (pseudoscalar) lattice-polariton or exciton-polariton modes in dielectric crystals permits us to observe the axion - photon and photon axion conversion processes by means of registration of emission due to absorption and luminescence processes, The same properties have the crystals ($BaWO_4$, $BaMoO_4$, $PbWO_4$, $PbMoO_4$), described by C_{4h}, C_{6h}, and others point symmetry groups. The experimental scheme for observation of resonant axion-polariton conversion processes is illustrated by Fig.1a. Dark points correspond to high energy axions, generated by Stars, Sun or dark matter of Universe. After absorption of axions in crystals as a result of axion - pseudoscalar polariton resonant interaction the luminescence emission may be detected by receiver 2 (see Fig. 1a). Now there are known the luminescent fibers, consisted from the low symmetrical scintillating microcrystals, having the pseudoscalar polar modes, active for one photon absorption or emission processes. So the experimental schemes for resonant axion – pseudoscalar polariton conversion processes with long size fibers may be also created (see Fig.1b). In such type experimental scheme the additional (outer) magnetic field also may be applied (see Fig.1b). At the case of sodium nitrite and some amino acid crystals (C_{2v} point group) pseudo scalar modes are not polar are allowed for two-photon (Raman) processes, but forbidden for one–photon emission (IR or UV). In these type structures pseudo scalar modes should have high Q-factor and high intensity in spontaneous Raman

Scattering. As the result there is the opportunity to observe the photon axion conversion processes by means of pseudoscalar modes excitation with the help of Stimulated Raman Scattering (SRS) or Two-Photon Absorption and registration of Two-Photon Emission (TPE) signal by detector 10 (see Fig.2a).

Table 1. Characters of presentations and selection rules for one- and two-photon processes of C_{2h}-point group; a_i and a_{ij} – are vector and the second order tensor components correspondingly; ps-pseudoscalar presentations components; V- vector presentation; $[V]^2$ – the presentation of symmetrical squared of the second order tensor.

C_{2h}	E	$C_2(z)$	σ_h	I	a_i	a_{ij}	ps
A_g	1	1	1	1		xx,yy,zz,xy	-
A_u	1	1	-1	-1	z		+
B_g	1	-1	-1	1		xz,yz	-
B_u	1	-1	1	-1	x,y		-
V	3	-1	1	-3			
$[V]^2$	6	2	2	6			

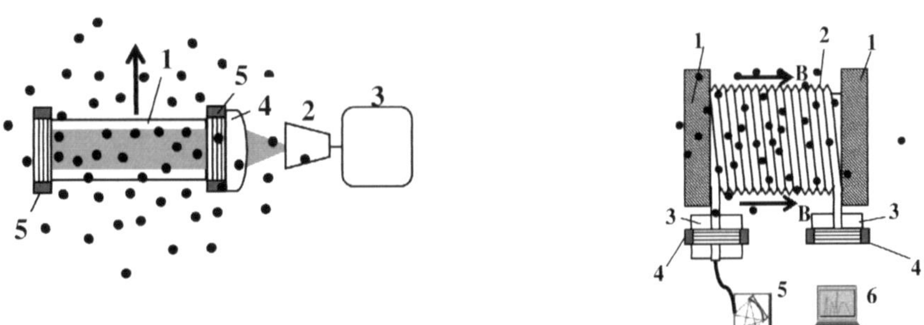

a b

Figure 1. Experimental schemes for observations of resonance axion-polariton conversion processes in dielectric crystals(a) and fibers(b);(a)- 1- monocrystals; 2-detector of infrared or ultraviolet (luminescent emission) emission; 3-computer; 4-lens; 5-mirrors; arrow show the direction of crystalline or outer magnetic fields;(b)-1,3-fixers; 2-luminescent fibers;4-mirrors;5-detector;6-computer.

Fig.2b illustrates the setup for observing "Light shining through wall" - Primakov-effect in dielectric crystals, having pseudoscalar polar excitons. Such properties are revealed by stilbene, antracene, naphthalene, POPOP, PPO, BaWO4,BaMoO4, PbWO4, PbMoO4 , Tb(NO3)3x5H20 and others crystals. In this case the UV laser source (1) excite the photoluminescence in scintillating crystal 5 with emitting of polar pseudoscalar excitons (polaritons), resonantly interacting with the corresponding axions (dark points at Fig.2). Axions penetrate through the wall 9 and result in the resemble photoluminescence in crystal 6. Spectrometers 7 should detect the luminescence emission from both crystals 5 and 6. The signal of the luminescent emission from crystal 6 is waited enough strong because of resonant character of axion–polariton interaction and due to the small group velocity of exciton polaritons with small wave vector.

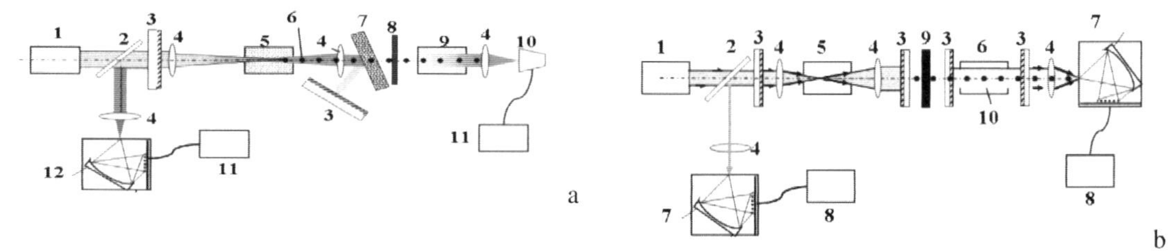

Figure 2. Experimental schemes for observation of one-(a) and two-(b)photon-axion conversion processes with realizing of "Light shining through wall" -Primakov-effect; (a)-1-high power pulsed laser source; 2-semitrasparent plate; 3-dielectric mirrors; 4-lens; 5-crystal for SRS excitation;6-SRS emission;7-photonic crystal; 8-opaque wall; 9- crystal for TPE emission;10 –detector of IR photons; 11-computers; 12-spectrometer for Raman spectra recording; (b)- 1- UV laser source; 2-semitrasparent plate; 3-dielectric mirrors; 4-lens; 5-crystal for luminescence excitation;6-crystal for luminescence emission;7-spectrometer; 8-computer; 9- opaque wall;10 –the secondary emission after wall.

Thus it is proposed the optimization of experimental installations for the detection of axion-photon conversion processes by means of using of dielectric crystals, having lattice or exciton pseudoscalar modes. Such crystal has low symmetry, described by point group C_{2h}, C_{4h}, C_{6h}, C_{2v} and others. For the first group of such crystals pseudoscalar modes have the same symmetry race as polar one. In this case axion-polariton conversion processes are permitted by selection rules without application of outer magnetic field. The using of scintillating fibers with resemble symmetry properties gives the opportunity to essentially increase the length of photon-axion interaction. For the generation of "hot" axions it is proposed to use laser for exciting of SRS and resonance photoluminescence on pseudoscalar modes in low symmetrical crystals. Thus the proposal to observe axion-photon processes in low symmetrical dielectric crystals opens the new opportunities for generation and detection axions in laboratory.

Acknowledgements

The work was supported by the Russian Foundation for Basic Research (projects 15-02-02882, 14-02-90406 и 14-02-00190) and Education and Science Department, Program "5 top 100".

References

[1] Ryabov V A, Tsarev V A, Tskhovrebov A M 2008 *Phys.Usp.* **51** 1091–121
[2] Appelquisi T, Cheng H C, Dobrescu B A 2001 *Phys.Rev. D* **62** 035002
[3] Servant G, Tait T M 2001 *Nucl. Phys. B* **650** p 391
[4] Goldstone J, Salam A, Weinberg S 1962 *Phys. Rev.* **127** p 965
[5] Kim J E, Garosi G 2010 *Reviews of Modern Physics* **82** pp 557-601
[6] Hoffmann S 1986 *Phys. Lett. B* **193** 117-22

[7] Picciotto C, Pospelov M 2005 *Phys. Lett. B* **605** p 15
[8] Okun L B, 1982 *Zh. Eksp. Teor. Fiz.* **83** 892-8
[9] Jaeckel J, Redondo J, Ringwald A 2009 *EPL* **87** 10010
[10] Sikivie P, Tanner D B and van Bibber K 2007 *Phys. Rev. Lett.* **98** 172002
[11] van Bibber K A, Dagdeviren N R, Koonin S E, Kerman A K, Nelson H N 1987 *Phys. Rev. Lett.* **59** 759-62
[12] Duffy L D, Sikivie P, Tanner D B, Asztalos S J, Hagmann C, Kinion D, Rosenberg L J, van Bibber K, Yu D B, Bradley R F 2006 *Phys. Rev. D* **74** 012006
[13] Rosenberg I J, Bibber K A 2000 *Phys. Rep.* **325** 1
[14] Stancil D D 2007 *Phys.Rev. D* **76** 111701(R)
[15] Afanasev A, Baker O K, Beard K B, Biallas G, Boyce J, Minarni M, Ramdon R, Shinn M, Slocum P 2008 *Phys. Rev. Lett.* **101** 120401
[16] Mueller G, Sikivie P, Tanner D B, van Bibber K 2009 *Phys. Rev. D* **80** 072004
[17] Hoskins J, Hwang J, Sikivie P, Tanner D B 2010 *Phys. Rev. Lett.* **104** 041301
[18] Friedland A, Giannotti M and Wise M 2013 *Phys. Rev. Lett.* **110** 061101
[19] Bellini G, Benziger J, Bick D et al 2012 *Phys. Rev. D* **85** 092003
[20] Hudson H S, Acton L W, DeLuca E, Hannah I G, Reardon K, Van Bibber K *arXiv*:1201.4607v1
[21] Gorelik V S and Izmailov G N 2011 *Bull. Lebedev Phys. Inst.* **38** 177–83
[22] Gorelik V S, Pyatyshev A Yu 2014 *Bull. Lebedev Phys. Inst.* **41** 316-22
[23] Gorelik V S 2015 *Herald of Bauman Moscow State Techninical University. Natural Sciences* **1** pp 35-54
[24] Gorelik V S 2015 *Bull. Lebedev Phys. Inst.* **42** pp 55-62
[25] Gorelik V S 2015 *EPJ Web of Conferences* **103** 04003
[26] Gorelik V S 2015 *Herald of Bauman Moscow State Techninical University. Natural Sciences* **4** pp 23-35
[27] Gorelik V S, Burdanova M G 2015 *Herald of Bauman Moscow State Techninical University. Natural Sciences* **5** pp 3-11
[28] Gorelik V S, Pyatyshev A Yu, Krylov A S 2016 *Physics of Solid State* **58** 170-6
[29] Gorelik V S, Burdanova M G 2016 *Laser Physics* **26** 035001
[30] Gorelik V S, Pyatyshev A Yu, Belyanchikov M A, Gorshunov B P 2016 *Physics of Wave Phenomena* **24** pp 87–95

On propagation of electromagnetic and gravitational waves in the expanding Universe

V O Gladyshev
Department of Physics, Bauman Moscow State Technical University, Moscow, Russia
E-mail: vgladyshev@mail.ru

Abstract. The purpose of this study was to obtain an equation for the propagation time of electromagnetic and gravitational waves in the expanding Universe. The velocity of electromagnetic waves propagation depends on the velocity of the interstellar medium in the observer's frame of reference. Gravitational radiation interacts weakly with the substance, so electromagnetic and gravitational waves propagate from a remote astrophysical object to the terrestrial observer at different time. Gravitational waves registration enables the inverse problem solution - by the difference in arrival time of electromagnetic and gravitational-wave signal, we can determine the characteristics of the emitting area of the astrophysical object.

The propagation time of electromagnetic radiation depends on the result of superposition of the source electromagnetic wave and secondary waves arising from the interaction between the primary wave and moving atoms of the medium.

Moving off the emitting astrophysical object from the observer leads to Doppler shift of frequency radiation in an expanding area, which is related to the movement of the space area where the object is located [1].

The spread of electromagnetic radiation in the interstellar medium of the expanding Universe will lead to a further time delay in light signals propagation.

We obtain an equation for electromagnetic wave propagation taking into consideration movements of the interstellar medium. Consider Robertson-Walker metric [2]

$$d\tau^2 = dt^2 - R^2(t)\left\{\frac{dr^2}{1-kr^2} + r^2 d\theta^2 + r^2 sin^2\theta d\varphi^2\right\}. \quad (1)$$

Here r, θ, φ, t are the coordinates, $R(t)$ is the cosmological scale factor, k is the spatial curvature, $R(t) = R_0/c$, R_0 is the common distance.

The equation for propagation of electromagnetic and gravitational-wave signal along the radial direction can be obtained from (1) when $d\theta = d\varphi = 0$

$$d\tau^2 = dt^2 - R^2(t)\left\{\frac{dr^2}{1-kr^2}\right\} = 0. \quad (2)$$

The expression for the invariant proper time is given by [3]

$$d\tau^2 = (1-\beta_e^2)dt^2, \quad (3)$$

where $\beta_e = v/c$, v is the group velocity of electromagnetic waves, c is the speed of light in vacuum.

If we equate this expression to the element of Robertson-Walker length (2), we obtain

$$\int_{t_1}^{t_0} \frac{dt}{R(t)} = \int_0^{r_1} \frac{1}{\beta_e(r,t)} \frac{dr}{\sqrt{1-kr^2}}. \quad (4)$$

Here t_1, t_0 are the moments of radiation and registration time, measured from the singular state, r_1 is dimensionless distance to the source of space radiation in the Earth's ISO. Before the integral on the right, the positive sign is chosen, which corresponds to the phase of the Universe expansion.

Since the cosmological expansion can be slowed down with time, the group velocity of the electromagnetic wave depends on t.

The dependence of the relative velocity of the interstellar medium on the coordinate along the light propagation path concerning the terrestrial observer, has the form

$$\beta_n(r,t) = \frac{\dot{R}(t)}{cR(t)}(r - r_1), \tag{5}$$

where r is the current radial coordinate. The parameter R with $k = 1$ can be characterized as the Universe radius. In the point of time $t = 0$ the radius $R = 0$, so the current time t_0 is the time measured from this singularity, and it can be called the age of the Universe.

The relative velocity of moving off the emitting astrophysical object without taking into account the speed of the object in the local area of expanding space equals

$$\beta(t) = -r_1\frac{\dot{R}(t)}{cR(t)}. \tag{6}$$

In the expanding Universe any kind of $R(t)$ at the expansion stage will increase the difference $t_0 - t_1$. This will lead to the same time delay of the light and gravitational signal arrival.

To estimate the effect of the interstellar medium motion, let us consider the properties of the integral on the right in (4).

Consider i - medium layer when the motion of the interstellar medium is directed opposite the wave vector of the electromagnetic wave. The phase velocity of radiation propagation in a moving medium can be found from the solution of the dispersion equation [4]. We introduce the following symbols:

$$y_i^{-2} = 1 - \beta_{in}^2, \qquad k_i = \varepsilon_i\mu_i - 1, \qquad \beta_{in} = \frac{u_{in}}{c}, \qquad \beta = \frac{V}{c}.$$

Here, the values u_{in} characterize normal and tangential velocity components, ε_i, μ_i are dielectric and magnetic permeability of i layer in the observer's ISO, V is the velocity of the radiation source motion in the observer's ISO, c is the velocity of a plane monochromatic electromagnetic wave in vacuum.

Consider the case when $\beta = 0$ for i - layer of the medium, when the interstellar medium motion is directed opposite the wave vector of the electromagnetic wave. For a terrestrial observer the phase velocity of electromagnetic wave propagation in the i - moving medium layer corresponds to

$$v_i = c\frac{1 - k_i\gamma_i^2\beta_{in}^2}{-k_i\gamma_i^2\beta_{in} + \sqrt{1 + k_i}}. \tag{7}$$

Assuming the function $v(r)$ to be continues for the signal propagation time, from (4) we obtain

$$\int_{t_1}^{t_0}\frac{dt}{R(t)} = \int_0^{r_1}\frac{-k_iy_i^2\beta_{in} + \sqrt{1 + k_i}}{1 - k_iy_i^2\beta_{in}^2}\frac{dr}{\sqrt{1 - kr^2}}. \tag{8}$$

As the interaction of gravitational radiation with the medium is negligibly small, the propagation time of a gravitational-wave signal will also be determined by (8), but when $\beta_e(r,t) = 1$.

Registration of relic cosmological gravitational waves would make it possible to determine the dependence of the scale factor on the time $R(t)$ [5].

Analysis of the integral equation kernel shows that the interstellar medium and the emitting area medium of the star can make a substantial contribution to the effect of time delay of light propagation.

We can also take into account that $n = \sqrt{\varepsilon\mu} = n(r,t)$ is the refractive index of the intergalactic medium and it depends on t, as in the expanding Universe the medium density changes with time.

Registration of gravitational waves could help to solve the inverse problem - according to the time difference in arrival of the electromagnetic and gravitational-wave signal, we can determine the characteristics of the emitting area of the astrophysical object.

Let us deal with the experiment when we measure the registration time of SN1987A burst of radiation by means of neutrino and gravitational-wave detectors. With the help of spaced detectors [6, 7] used in this experiment, we measured an abnormally long signal registration time delay. The burst was registered by the gravitational antennas in Maryland and Rome, as well as by the neutrino detector in Monte Blanco, which is bound to Greenwich Mean Time. Detector reading is correlated for 2 hours with a 1.1 second signal lag recorded by the neutrino detector.

The measured registration time delay of a signal propagating with the speed of light in a vacuum in any ISO is a consequence of different propagation velocity of the neutrino and gravitational signals. From the above analysis, we make a conclusion that the determination of the cosmological distance must take into account the effect of the light propagation delay in a moving medium.

References

[1] Harrison E R 1981 Cosmology, the Science of the Universe. *Cambridge U. P.* pp 216-218

[2] Tolman R C 1987 Relativity, thermodynamics, and cosmology. *Oxford at the Clarendon Press* pp 501

[3] Weinberg S 1972 *Gravitation and Cosmology: Principles and Applications of the General Theory of Relativity*. p 657 (New York: Willey)

[4] Gladyshev V, Gladysheva T, Zubarev V 2006 Propagation of electromagnetic waves in complex motion media. *Journal of Engineering Mathematics* **55** no. 1-4 pp 239-254

[5] Gladyshev V O, Morozov A N 2000 Classification of Gravitational-Wave Antennas by the Methods of Gravitational Radiation Detection. *Measurement Techniques* **43** no. 9 pp 741-746

[6] Pizzella G 1990 Correlations among gravitational wave and neutrino detector date during SN1987A. *Nuovo cim. B* **105** no. 8-9 pp 993-1008

[7] Pizzella G 1992 Correlations between gravitational-wave detectors and particle detectors during SN1987A. *Nuovo cim. C* **15** no. 6 pp 931-941

Journal of Physics: Conference Series **731** (2016) 012009 doi:10.1088/1742-6596/731/1/012009

K.P. Stanyukovich and GR Extensions by his colleagues in Diverse Dimensions

V N Melnikov
Head of Center for Gravitation and Fundamental Metrology, VNIIMS, Moscow, Russia
Professor, Institute of Gravitation and Cosmology, Peoples' Friendship University of Russia, Moscow, Russia
E-mail: melnikov@vniims.ru

Abstract. Main stages and scientific results of K.P. Stanyukovich and also of his colleagues after 1989 in gravitation, cosmology and astrophysics are described both in four and arbitrary dimensions.

The 3rd of March, 2016, was the 100th birthday of Professor Staniukovich (1916-1989) – the outstanding Russian scientist, who contributed significantly to the development of gas dynamics, physics of explosion, magnetic hydrodynamics, astronomy and theory of gravitation. He was involved in making new scientific trends such as cumulation theory, strong explosion theory, magnetic and cosmic gas dynamics, theory of self-similar solutions to differential equations and cosmoparticle physics.

On March 3-4, 2016, the traditional International conference on general relativity, cosmology, relativistic astrophysics and non-stationary gas dynamics, devoted to his jubilee, was held in the Bauman Moscow State Technological University (BMSTU), Russia. The conference program included reports and informal discussions on the latest developments in the fields of main interests of Prof. Staniukovich, such as general relativity extensions, theoretical and observational cosmology and astrophysics, gravitational experiments, fundamental constants, gas dynamics, explosion theory etc. Similar very successful conferences devoted to his 80th and 90th jubilees were held in Moscow at the Peoples' Friendship University of Russia (PFUR) in 1996 and 2006.

This third one was organized by BMSTU, the Russian Gravitational Society, Centre for Gravitation and Fundamental Metrology of VNIIMS and Institute of Gravitation and Cosmology, PFUR and Institute of Natural Science and Technique History, RAS.

Many Russian scientists in the above-mentioned fields took part in the conference. They represented groups from Moscow: (Lomonosov Moscow State University, BMSTU, PFUR, Center for Gravitation and Fundamental Metrology of VNIIMS, Lebedev Physical Institute, RAS, Kurchatov Atomic Energy Institute, Nuclear Safety Institute, Physical-Technical Institute, National Nuclear Research Institute, Moscow State Pedagogical University etc.; from St.-Petersbourg: SPb State University, SPb Financial Academy, Friedman Lab.; from Kazan: Kazan State Federal University; from Tomsk: State and Pedagogical Universities: from Ulyanovsk Pedagogical University and Perm State University as well as scientists from NIS. More than 150 participants presented plenary and sectional talks. Some of them are represented in this book.

First we wish to remind here the main facts of Prof. Staniukovich's scientific biography.

His first scientific paper (in astronomy) appeared in 1930 when he was only 14 years old. Other more serious works, since 1932, were related to meteoric physics, theory of meteoric origin of lunar craters and theory of shocks with cosmic velocities. The latter, which he proposed in 1937, is topical for modern problems of astronautics as well. His works on meteoric shocks are used to study the meteoric danger in space flights.

During the Great Patriotic War (1941-1945), K.P. Staniukovich successfully worked on the instructions of the State Defense Committee.

In 1944, in collaboration with L.D. Landau, he solved the problem of the converging strong shock wave, being of importance for theoretical physics and practical implementation. This result promoted solving some problems since super-high pressures are achieved in the shock waves. It should be noted that very few people could manage to publish a joint paper with Landau, but K.P. Staniukovich published with him 3 papers!

In 1944-1946 K.P. Staniukovich obtained fundamental results in the field of detonation theory and propagation of blast waves. He elaborated theories describing detonation of condensed explosives, cumulation of a compressible medium, body throwing by explosion products being in the wave process. In 1944 K.P. Staniukovich created the strong explosion theory, which permitted one to improve the utilization factor and to calculate the damage effect of a shock wave.

Of utmost importance were K.P. Staniukovich's works on the "thermal death" of the Universe. In 1949, K.P. Staniukovich published the article "On Entropy Increase in the Infinite Universe", where he showed the impossibility of a thermal death of the Universe This paper was recommended for publication by the president of USSR Academy of Science S.I. Vavilov. It was repeatedly cited in this country as well as abroad. K.P. Staniukovich is the author of pioneering works on non-relativistic and relativistic magnetic hydrodynamics.

K.P. Staniukovich's work on unsteady motions of a continuous medium considerably promoted the development of many branches of technology. They found application in explosive treatment of materials and are used in mining industry and civil engineering. He was awarded the USSR State Prize for his contribution in these fields in 1981.

Since 1961 K.P. Staniukovich successfully worked at the problems of gravity theory and cosmology, their connection with particle physics and problems of possible variations of fundamental physical constants [1, 2]. The second half of the 20th century in the field of gravitation was devoted mainly to theoretical study and experimental verification of general relativity and alternative theories of gravitation with a strong stress on relations between macro and micro-world phenomena or, in other words, between classical gravitation and quantum physics. Very intensive investigations in these fields were done in Russia by K.P. Staniukovich and M.A. Markov from mid 60's, then by Ya.B. Zeldovich, A.D. Sakharov with their colleagues and many others.

K.P. Staniukovich elaborated a hydrodynamic model of the gravity "mechanism", versions of the gravity theory with a variable number of particles, proposed the first scenarios of evolution of the Universe with an initial vacuum stage and the decay of super heavy particles, obtained relationships for characteristic lengths and masses of various fundamental interactions, which was an important step on the path of creating a unified theory of all physical interactions [1, 2] etc. His last work was devoted to new solutions of nonlinear difference equations. Once N.N. Bogolyubov told his own colleagues and students at the informal discussion in the Moscow State University: "if you can not solve a nonlinear equation, go for help to Kirill Petrovich Stanyukovich, he will do it, or it is not solvable now".

K.P. Staniukovich's scientific activity is widely known in the country as well as abroad. He is the author of 27 monographs and booklets on modern problems of physics and mechanics. He wrote more than 430 scientific and popular-science works and has more than 20 author certificates, including the book "Unsteady Motions of a Continuous Medium", Pergamon Press) which have been translated into foreign languages and are universally recognized.

A high skill combined with great pedagogical activity allowed Prof. K.P. Staniukovich to create scientific schools of specialists in gas dynamics and theoretical physics. In 1953-1958 he headed the Higher Mathematics Department at the Bauman Moscow Higher Technical School (now Technical University).

He delivered courses of lectures in gas dynamics and explosion physics in BMHTS and of magnetic hydrodynamics and theory of gravitation at the N.N. Bogolyubov' chair of the Department of Physics at Moscow State University.

Among his disciples are more than 25 D.Sci.'s and 50 Ph.D.'s who were and are working in Moscow, Kirghizia, Kazakhstan, Armenia and Georgia (Tbilisi).

For many years K.P. Staniukovich had been the Editor-in-Chief of the annual "Problems of Gravitation Theory and Particle Theory", which played a crucial role in the development of new trends in gravitation theory and especially cosmoparticle physics and which is essentially the predecessor of the international journal "Gravitation and Cosmology".

But the main achievement of Prof. Staniukovich was the foundation in 1967 of a large group of young scientists effectively working in the fields of gravitation, cosmology, cosmoparticle physics and theory of gravitational experiment in the Russian State Committee for Standards. Many of them became leading Russian scientists in these fields. One should mention the fundamental works of G.A. Vilkovysky on gravitational field quantization, M.B. Mensky on groups theory and continuous measurements, V.A. Belinsky on the general solution of Einstein equations near the singularity, G.A. Sokolik and N.P. Konopleva on gauge fields theory, Yu.N. Barabanenkov on statistical physics and quantum cosmology, V. Yakimov on ASTRON Project, G.N. Shikin on exact solutions with nonlinear fields in gravity, O. Karageoz on very precise absolute G measurements.

Also, the author and K.A. Bronnikov with colleagues and their students - on exact solutions of Einstein equations with scalar and electromagnetic fields and their interactions, the first quantum cosmological model with a cosmological constant Melnikov and Kalinin (creation from nothing) (1972); first classical cosmological models for conformal scalar field (1968) and quantum cosmological models with minimal and conformal scalar fields (Melnikov and Reshetov, 1971), first nonsingular cosmological model with spontaneous symmetry breaking (Melnikov and Orlov, 1978-79) of the nonlinear conformal scalar field, exact solutions for nonlinear electrodynamics, including Born-Infeld one, first exact solution for dilaton-type interaction with electro-magnetic field in GR, first non-singular field particle-like solution with gravitational field (Bronnikob, Melnikov, Shikin and Stanyukovich, 1979). Also, the conclusion that only G may vary with respect to atomic system of measurements in the Jordan frame (Zaitsev and Melnikov, 1978). See realization of all this program in [1].

Later, from 1987 in the frames of multidimensional Einstein equations, which are the low energy limit of unified models, the author mainly with V.D. Ivashchuk and also with V.R. Gavrilov and K.A. Bronnikov with students and post-graduates using mainly sigma model approach were realizing the second program in arbitrary dimensions in cosmology and gravitation [3-6] obtaining exact solutions with different sources: with a cosmological constant , also with a scalar field (we singled out nonsingular, dynamically compactified, inflationary solutions, 1994); a perfect fluid (PF), a PF with a scalar field (e.g., nonsingular, inflationary solutions); a viscous fluid (e.g., nonsingular solutions with generation of mass and entropy, quintessence and coincidence in a 2-component model, Melnikov and Gavrilov); oscillating behavior near the singularity, billiards in Lobachevsky space for systems with different sources (Ivashchuk and Melnikov). In all cases mentioned above, solutions with Ricci-flat factor spaces were obtained for any n, also with curvature in one factor space; with curvatures in 2 factor spaces - only for the total dimensions $D = 10; 11$; with fields: scalar and forms of arbitrary rank (Ivashchuk and Melnikov, 1998); inflationary solutions generated by forms (p-branes); first billiards for models with branes (Ivashchuk and Melnikov, 1999); quantum generalization (solutions to WDW-equation) for certain cases where classical solutions were obtained; solutions for models with dilatonic fields with potentials, billiard behavior for them also and recently quantum billiards for D-dimensional cosmological models of generalized Bianchi-I type (all this with Ivashchuk). Some of these integrable models were also used for calculation of temporal and space variations of the effective gravitational and fine structure constants, including the case with nonlinear in curvature model (with Bronnikov). Comparison with the present experimental bounds allowed us to choose particular models or to single out some classes of solutions.

A class of solutions depending on the radial variable in any dimensions was found as well (with Ivashchuk and Bronnikov): generalized Schwarzchild, Tangherlini solutions where black holes (BHs) were singled out, also with a minimal scalar field '; generalized Reissner-Nordstrom solutions (BHs were also singled out), the same was done with scalar field (no BHs); multi-temporal solutions; for dilaton-like interaction of scalar and electromagnetic fields (BHs exist only for a special case); stability studies (stable solutions only for the previous case); for obtained dilaton-forms (p-brane) interaction, the solutions' stability was found only in some cases, e.g. for one-form. PPN parameters for most models were calculated.Some new results on constants their variations and quantum billiards see in [7-9].

After the discovery of the dark energy and dark matter we see that our present knowledge of the Universe is very poor but we still may hope that the possible discoveries of the Higgs boson and gravitational waves will help us to move forward in the endless way of understanding the Nature may be through the better understanding the nature of gravitation as a missing link to unification of known physical interactions !

I want to finish these short remarks saying that K.P. Staniukovich was an open-hearted person, with a good sense of humour, who maintained a pleasant creative atmosphere in the research group he headed for about three decades. It was always a real pleasure to work and contact with him!

Acknowledgments

The work is partially supported by the RFBR grant 16-02-00602a and by the Ministry of Education and Science of the Russian Federation on the Program to improve the competiveness of the People's Friendship University of Russia among the world's leading research and education centers in 2016-2020.

References

[1] Staniukovich K P 1965 *Gravitational Field and Elementary Particles* (Moscow : Nauka)

[2] Staniukovich K P and Melnikov V N 1983 *Hydrodynamics, Fields and Constants in the Theory of Gravitation* (Moscow : Energoatomizdat)

[3] Melnikov V N 1993 *Exact Solutions and Variations of Constants* (Rio de Janeiro : CBPF) NF-051/93

[4] Melnikov V N 1995 *Multidimensional Cosmology and Gravitation* (Rio de Janeiro: CBPF) MO-002/95

[5] Melnikov V N 2002 *Exact Solutions in Multidimensional Gravity and Cosmology III* (Rio de Janeiro: CBPF) MO-03/02

[6] Melnikov V N 2006 *Proc. Albert Einstein Century International Conference (Paris)* **861** (New York : AIP Conf Proc) p 109

[7] Bronnikov K A and Melnikov V N et al 2013 *Gen. Rel. Gravitation.* **45** 12 2509

[8] Ivashchuk V D and Melnikov V N 2014 *Euro Phys. J* **3** 2805 1

[9] Ivashchuk V D and Melnikov V N 2016 *J. Mod. Phys.* A **31** 1641009

Space-time with a fluctuating metric tensor model

A N Morozov
Head of "Physics" Department, Bauman Moscow State Technical University, Moscow, Russia
E-mail: amor59@mail.ru

Abstract. A presented physical time model is based on the assumption that time is a random Poisson process, the intensity of which depends on natural irreversible processes. The introduction of metric tensor space-time fluctuations allowing describing the impact of stochastic gravitational background has been demonstrated. The use of spectral lines broadening measurement for the registration of relic gravitational waves has been suggested.

1. Introduction

A fluctuating physical time model was first proposed in the work [1] to explain the features of ion mobility fluctuations in a small amount of electrolyte under review [2, 3]. Based on the equation binding the entropy production density and the intensity of physical time fluctuations, the work [4] calculates the intensity of fluctuations specified and compares with experimental results.

2. Fluctuating physical time model

Let us consider the fluctuating physical time model, based on the assumption that the physical time under review is a random Poisson process $\tau(t)$ with fluctuations, equal to $\tau_0 = 1/\nu_\tau$, where ν_τ - the intensity of physical time fluctuations.

The one-dimensional characteristic function of a Poisson process under review is given by [5, 6]

$$g_\tau(\mu_\tau;t) = \exp((\exp(i\tau_0\mu_\tau)-1)\nu_\tau t), \tag{1}$$

and its n-dimensional characteristic function at $t_n \geq \ldots \geq t_2 \geq t_1$ respectively takes the form

$$g_{n\tau}(\mu_{1\tau},\mu_{2\tau},\ldots,\mu_{n\tau};t_1,t_2,\ldots,t_n) = \exp\left(\nu_\tau \sum_{j=1}^{n} t_j \left(\exp\left(i\tau_0 \sum_{k=j}^{n} \mu_{k\tau}\right)-1\right)\right). \tag{2}$$

The presented characteristic functions (1) and (2) make it possible to register the expectation function $\langle\tau(t)\rangle$ and correlation function $\langle\tau(t_1)\tau(t_2)\rangle$ for arbitrary time points t_1 and t_2 as follows:

$$\langle\tau(t)\rangle = t, \tag{3}$$

$$\langle\tau(t_1)\tau(t_2)\rangle = t_1 t_2 + \tau_0 \min(t_1,t_2). \tag{4}$$

Let us introduce the function

$$\xi_T(t) = \frac{d\tau(t)}{dt}, \tag{5}$$

expectation function $\langle\xi_T(t)\rangle$ and correlation function $\langle\xi_T(t_1)\xi_T(t_2)\rangle$ of which are given by

$$\langle\xi_T(t)\rangle = 1, \tag{6}$$

$$\langle\xi_T(t_1)\xi_T(t_2)\rangle = \tau_0\delta(t_2-t_1). \tag{7}$$

The introduction of the function $\xi_T(t)$ allows presenting the differential $d\tau(t)$ in the form of

$$d\tau(t) = \xi_T(t)dt. \tag{8}$$

Please note that the same expression is used in the general relativity theory to establish the relation between proper time τ and time coordinate $x^0 = ct$ [7]

$$d\tau = \frac{1}{c}\sqrt{g_{00}}dx^0, \tag{9}$$

where c - speed of light in vacuum, g_{00} - metric tensor component g_{ik}, $i,k = \overline{0,3}$, describing the space-time time coordinate variation. The comparison of formulas (8) and (9) allows presenting the value g_{00} as follows:

$$g_{00} = \xi_T^2, \tag{10}$$

which indicates a fundamental possibility of making a more general description of a space-time with a fluctuating metric.

3. Space-time with a fluctuating metric tensor model

The statement of the problem of physical processes description within a space-time model with a randomly fluctuating metric tensor is associated not only with the fluctuating physical time model, but also with the possibility of potential fluctuating space curvature due to the composition of a large number of gravitational waves [8]. At the same time, considering gravitational waves generating by different, uncorrelated against each other astrophysical objects such as neutron stars, black holes etc., as well as occurring as a result of gravitational collapse and supernova outburst, the model of randomly fluctuating metric tensor for space-time description is quite reasonable. Moreover, the model proposed may be used to describe the stochastic background caused, in particular, by a relict gravitational radiation [9, 10].

In general, the statement of the problem of physical processes description in the space-time with a fluctuating metric tensor model can be formulated as follows. Let us assume that any physical process, for example, electromagnetic wave propagation, proceeds in a space-time model. Let us consider that the metric tensor $g_{ik}(x^0, x^1, x^2, x^3)$ constitutes a random coordinate function $x^l = \{x^0, x^1, x^2, x^3\}$. It is required to determine the history of the physical process under review in case of metric tensor $g_{ik}(x^l)$ fluctuating randomly with time and space shift.

To create equations describing the physical processes proceeding in the space-time with a fluctuating metric tensor model, the method, described in the work [6], can be applied. Electromagnetic field equations in a curved space-time allow analyzing, in particular, the character of electromagnetic waves propagation in the space-time with a fluctuating metric tensor model [11].

4. Light propagation in the space-time with a fluctuating metric tensor model

Let us further consider the case of light propagation description with geometrical optics approximation. This case may occur if short light wavelength λ_0 condition is met, in comparison to the characteristic correlation length δL of the metric tensor fluctuations: $\lambda_0 << \delta L$. The equation is describing the light propagation in a curved space-time with geometrical optics approximation is given by [7]:

$$\frac{dk^i}{d\chi}+\Gamma^i_{kl}k^k k^l=0, \tag{11}$$

where: k^i - four-dimensional wave vector, χ - parameter, fluctuating along a light ray, Γ^i_{kl} - Christoffel symbols, expressed through the metric tensor as follows:

$$\Gamma^i_{kl}=\frac{1}{2}g^{im}\left(\frac{\partial g_{mk}}{\partial x^l}+\frac{\partial g_{ml}}{\partial x^k}-\frac{\partial g_{kl}}{\partial x^m}\right). \tag{12}$$

Formulas (11) and (12) and further assume the summation over repeated indices, and the metric tensor g_{ik} becomes symmetric: $g_{ik}=g_{ki}$.

Let us further consider the case when the metric tensor g_{ik} is considered little different from the Galilean metric $g^{(0)}_{ik}$:

$$g_{ik}=g^{(0)}_{ik}+h_{ik}, \tag{13}$$

where: $|h_{ik}(x^l)|<<1$, $g^{(0)}_{00}=1$, $g^{(0)}_{11}=g^{(0)}_{22}=g^{(0)}_{33}=-1$; $g^{(0)}_{ik}=0$ at $i\neq k$. Then, first approximation Christoffel symbols can be presented as follows:

$$\Gamma^i_{kl}=\frac{1}{2}g^{im}_{(0)}\left(\frac{\partial h_{mk}}{\partial x^l}+\frac{\partial h_{ml}}{\partial x^k}-\frac{\partial h_{kl}}{\partial x^m}\right), \tag{14}$$

where $g^{im}_{(0)}=g^{(0)}_{im}$.

Following the work [12], the solution of the equation (11) will be presented in expanded form:

$$k^i=k^i_{(0)}+k^i_{(1)}+\dots. \tag{15}$$

Then, zero-order approximation equation (11) takes the form:

$$\frac{dk^i_{(0)}}{d\chi}=0. \tag{16}$$

Therefore, $k^i_{(0)}=const$.

First approximation equation (11) integration results in

$$k^i(\chi)=k^i_{(0)}+k^i_{(1)}(0)-k^k_{(0)}k^l_{(0)}\int\limits_0^\chi \Gamma^i_{kl}d\chi. \tag{17}$$

The formula for the four-dimensional vector $k^i_{(1)}(0)$ can be derived from the condition [7]:

$$g_{ik}k^i k^k=0 \tag{18}$$

and in the first approximation is given by:

$$k^i_{(1)}=-\frac{1}{2}g^{im}_{(0)}h_{ml}(0)k^l_{(0)}. \tag{19}$$

Then, the solution (17) takes the final form:

$$k^i(\chi)=k^i_{(0)}-\frac{1}{2}g^{im}_{(0)}h_{ml}(0)k^i_{(0)}-k^k_{(0)}k^l_{(0)}\int\limits_0^\chi \Gamma^i_{kl}d\chi. \tag{20}$$

Since vector k^i can be presented as [7]

$$k^i = \frac{dx^i}{d\chi}, \tag{21}$$

in zero-order approximation it takes the form:

$$x^i(\chi) = k^i_{(0)}\chi + x^i(0). \tag{22}$$

Let us consider the case of light ray propagation in the direction of the axis x^1. Then we may assume that the components of vector $k^i_{(0)}$ have the following meanings: $k^0_{(0)} = k^1_{(0)} = k_0$, $k^2_{(0)} = k^3_{(0)} = 0$, where k_0 - wave number. Considering the distance l, passed by the light, equal to: $l = x^1(\chi) - x^1(0)$, from the formula (22) at $i = 1$, we obtain an expression for the parameter χ:

$$\chi = \frac{l}{k_0}. \tag{23}$$

Based on assumptions, from the formula (22) the following expressions were obtained:

$$\frac{k^\alpha(l)}{k_0} = 1 - \frac{1}{2}(-1)^\alpha (h_{0\alpha}(0) + h_{1\alpha}(0)) - \int_0^l \left(\Gamma^\alpha_{00} + 2\Gamma^\alpha_{01} + \Gamma^\alpha_{11}\right)dl,\ \alpha = 0,1, \tag{24}$$

$$\frac{k^\beta(l)}{k_0} = \frac{1}{2}\left(h_{0\beta}(0) + h_{1\beta}(0)\right) - \int_0^l \left(\Gamma^\beta_{00} + 2\Gamma^\beta_{01} + \Gamma^\beta_{11}\right)dl,\ \beta = 2,3, \tag{25}$$

where the Christoffel symbols $\Gamma^i_{kl} = \Gamma^i_{lk}$ symmetry is taken into account.

Formulas (24) and (25), after inserting an expression for the Christoffel symbols (14), allow to make an expression for relative frequency $\delta\omega$ and the wave vector δk^α fluctuations at the point of observation:

$$\frac{\delta\omega}{\omega_0} = \frac{\delta k^0}{k_0} = -\frac{1}{2}(h_{00}(0) + h_{01}(0)) + \int_0^l \left(\frac{1}{2}\frac{\partial(h_{11} - h_{00})}{\partial x^0} - \frac{\partial(h_{00} + h_{01})}{\partial x^1}\right)dl, \tag{26}$$

$$\begin{aligned}\frac{\delta k^\alpha}{k_0} &= \frac{1}{2}(h_{0\alpha}(0) + h_{1\alpha}(0)) + \\ &+ \int_0^l \left(\frac{\partial(h_{0\alpha} + h_{1\alpha})}{\partial x^0} + \frac{\partial(h_{0\alpha} + h_{1\alpha})}{\partial x^1} - \frac{1}{2}\frac{\partial(h_{00} + 2h_{01} + h_{11})}{\partial x^\alpha}\right)dl, \alpha = \overline{1,3}.\end{aligned} \tag{27}$$

Given that according to the formulas (22) and (23), we can state the following: $dx^1 = dl$, let us integrate the addend within the integral of the expression (26). Then we obtain the following:

$$\frac{\delta\omega}{\omega_0} = \frac{1}{2}(h_{00}(0) + h_{01}(0)) - h_{00}(l) - h_{01}(l) + \frac{1}{2}\int_0^l \frac{\partial(h_{11} - h_{00})}{\partial x^0} dl. \tag{28}$$

Let us introduce the tensor h_{ik} in the form of

$$h_{ik} = h\xi_{ik},\ i,k = 0,1, \tag{29}$$

Journal of Physics: Conference Series **731** (2016) 012010 doi:10.1088/1742-6596/731/1/012010

where: h - tensor h_{ik} fluctuations amplitude, $\xi_{ik} = \xi_{ik}\left(x^0, x^1\right)$ - random field with a unit amplitude and a correlation length $\delta L < \lambda_0$. In this case, all the addends preceding the integral within expression (28), will have the order h size.

Let us estimate the characteristic values of the integral within the expression (28). Since the geometrical optics approximation application is limited to the limit value of characteristic dimensions of correlation $\delta L \approx \lambda_0 = 1/k_0$ and characteristic time of correlation $\delta\tau \approx 1/\omega_0$, the frequency fluctuations variance estimate described by the integral in the expression (28), takes the following form:

$$\sigma_\omega^2 \approx h^2 l k_0 \tag{30}$$

or

$$h \approx \frac{\sigma_\omega}{\sqrt{l k_0}}. \tag{31}$$

In case of $l >> 1/k_0$, the integral contribution for the light frequency fluctuations becomes decisive.

Let us estimate the spectral line broadening when light passes through the space with a fluctuating metric. If it is assumed that the value l is approximately equal to the characteristic size of the space: $l \approx 10^{25}...10^{26}$ m (1...10 billion light years), $k_0 = 10^7$ m^{-1}, and the minimum registered spectral line broadening equals to $\sigma_\omega \approx 10^{-6}...10^{-8}$, the fluctuations amplitude h, calculated from the formula (31), takes the following form: $h \approx 10^{-22}...10^{-24}$.

Thus, when light propagates over a distance comparable to the size of the space, in the case of fluctuating metric with an amplitude $h \approx 10^{-22}...10^{-24}$ and frequency, upper-bounded with a light wave frequency, the spectral lines broadening shall be observed. The registration of an irreversible process of spectral lines broadening during the light propagation can evidence the existence of fluctuations of the space-time metric tensor and, consequently, the time fluctuations.

5. The calculation of the gravitational waves effect on light spectral lines broadening

If the description is carried out in a synchronous reference system [7], as usually done when calculating the plane gravitational waves effect within the linear theory framework [12], the Lorentz gage presents the following:

$$h_{00} = h_{0i} = 0. \tag{32}$$

Then the expression (28) for the light fluctuations relative frequency takes a simple form:

$$\frac{\delta\omega}{\omega_0} = \frac{1}{2}\int_0^l \frac{\partial h_{11}}{\partial x^0} dl. \tag{33}$$

Assuming that the gravitational waves filling the space have the frequency limit ω_g and a corresponding wave number k_g, the variance of the minimum detectable fluctuations of the space-time metric when measuring the light spectral lines broadening can be estimated by the following formula:

$$\sigma_h \approx \sqrt{\frac{c}{c\omega_g}}\sigma_\omega, \tag{34}$$

or based on measurement results averaging over a period of T :

$$\sigma_h \approx \frac{1}{\sqrt{T\omega_g}}\sqrt{\frac{c}{c\omega_g}}\sigma_\omega. \tag{35}$$

If $T = 3\cdot 10^7$ sec, then

$$\sigma_h \approx 10^{-31}...10^{-32}. \tag{36}$$

To calculate the gravitational waves $\Omega_{GW}(\omega)$ energy density in terms of their amplitude h, the formula given in the work [13] can be used:

$$\Omega_{GW}(\omega) = \frac{1}{6\pi H_0^2}\omega^3 G_h(\omega), \tag{37}$$

where H_0 - Hubble parameter. Then the relic gravitational waves $\omega_g = 10^9...10^{10}$ s^{-1} frequency limit formula is given by:

$$G_h(\omega) \approx 2\cdot\left(10^{-61}...10^{-64}\right)\Omega_{GW}(\omega) \tag{38}$$

or based on estimate of the value $\Omega_{GW}(\omega)$ on indicated frequencies within the quintessential inflation model framework [9]:

$$\sigma_h \approx 10^{-32}...10^{-33}. \tag{39}$$

The resulting value (39) is close to the minimum observed value (36).

6. Conclusion

Thus, the space-time with a fluctuating metric tensor model proposed allows describing potential physical time fluctuations. It was demonstrated that the measurement of the broadening of light spectral lines spreading over the space filled with high-frequency relic gravitational waves allows us to estimate the upper limit of their amplitude and possibly explore their spectral composition.

The author expresses his sincere gratitude to A.N. Nikitin for assistance in research and preparing the article for publication.

References

[1] Morozov A N and Nazolin A L 2001 *Dynamic systems with fluctuating time*. (Moscow: Bauman MSTU)
[2] Morozov A N 1997 *Irreversible processes and Brownian motion*. (Moscow: Bauman MSTU)
[3] Korotaev S M, Morozov A N, Serdyuk V O and Sorokin M O 2002 *Russian Physics Journal* vol **45** Iss 5 pp 431-444
[4] Morozov A N 2016 *Radiooptics of the Bauman MSTU* **01** pp 35-61
[5] Pugachev V S and Sinitsyn I N 1990 *Stochastic Differential Systems*. (Moscow: Nauka Publ.)
[6] Bunkin N F and Morozov A N 2011 *Stochastic Systems in Physics and Technology*. (Moscow: Bauman MSTU)
[7] Landau L D and Lifshitz E M 2003 *Field theory*. (Moscow: FIZMATLIT Publ.)
[8] Pustovoit V I, Morozov A N, Gladyshev V O, Ismaylov G N 2016 *Laser Gravity-Wave Antennas*. (Moscow: Bauman MSTU)
[9] Sa P M and Henriques A B 2010 *Phys. Rep. D.* **81** 124043
[10] Esakov A A, Morozov A N, Tabalin S E and Fomin I V 2016 *Herald of the Bauman Moscow State Tech. Univ., Nat. Sci.* **1** pp 26–35
[11] Morozov A N and Gladyshev V O 2002 *Russian Physics Journal* **45** (2) pp 113-117

Journal of Physics: Conference Series **731** (2016) 012010 doi:10.1088/1742-6596/731/1/012010

[12] Bichak I and Rudenko V N 1987 *Gravitational waves in GTR and problems of their detection* (Moscow: Bauman MSTU)
[13] Astone P et al 1990 *Astronomy and Astrophysics* **343** pp 19-22

On the direct detection of gravitational waves, and some of the problems of improving laser interferometers

V I Pustovoyt
Center "Photonics", Bauman Moscow State Technical University, Moscow, Russia
E-mail: vlad_pst@yahoo.com

Abstract. In this paper we describe an observational method for determining black holes masses. The study shows that the knowledge of the recorded low-frequency gravitational waves and the period from the beginning of registration till the moment of black holes collapse is sufficient and even preferable in determining the possible values of collapsing black holes masses. The reason for this is that the proportion of the period in the measured interval containing relativistic corrections (i.e. those ones in which the black hole speed is comparable to the speed of light), is smaller if the observed and measured time interval is longer. The values of black holes masses and the measured time interval, obtained as a result of the first observations, according to this model are in a very good agreement.We examine the problem of mirror heating in Fabry-Perot cavity of Michelson interferometer, by incident laser radiation, and we conclude that the problem of heat removal can be solved by a different approach, without use of multilayer reflective openings. As an alternative approach to the creation of highly reflective structures, we suggest using a spatially extended structure with a sinusoidal variation of the refractive index. We consider some of the possible technological methods for producing such structures based on heterogeneous media.The article describes the effects of the incident laser radiation exposure on the periodic structure, and it shows that the volume ponderomotive force may lead to a mirror polarization due to the radiation, and consequently, to appearance of an additional mechanical connection of the mirror with the surrounding mirror suspension design. The article examines the impact of the surface ponderomotive forces on the media boundary with different dielectric permeability and it shows that pressure spatial variables arising at the same time lead to deformation of the media layers, and the deformation and pressure values depend on the difference in the dielectric permeability values of the nearest layers. We also state the possible consequences of such spatially variable pressures effects (violation of synchronism conditions (misalignment) of reflective coatings, and others. It is also noted that our work, in which a laser interferometric technique for the detection of gravitational waves was first proposed, was already known to Joseph Weber in 1963.

The February issue of Physical Review Letters published an article which reported on the direct observation of gravitational waves from the confluence of two black holes [1]. This event is given the abbreviated name GW150914, which stands for: Gravitational Wave 2015, the 9th month, the 14 day, thus "pointing" on the principle of formation of such events in the future. The observation was carried out by means of the two unique independent interferometers LIGO: one in Hanford (Washington, USA), the other in Levingstoune (Louisiana, USA), the distance between them being 3002 km, so the time delay of the signal was about 7 ms, which corresponds to the speed of gravitational wave propagation at the speed of light. The peak amplitude of the gravitational wave (dimensionless value) was about 10^{-21}, which corresponds to a shift of the 4 kilometer arms of the interferometer. We proposed a fundamental scheme of the laser interferometer for the direct registration of gravitational waves as early as 1962 [2] and Rein

Journal of Physics: Conference Series **731** (2016) 012011 doi:10.1088/1742-6596/731/1/012011

Weiss supplemented it with his proposal to set Fabry-Perot cavity in each of the arms of the interferometer, which would extend the arms of the interferometer, increase the phase progression and thereby would improve the sensitivity [3]. The signal obtained with the interferometers, which is given in [1], is shown in Fig.1 (above). The figure shows that the signal is a "chirp" whose frequency ranges from 35Hz to 250Hz.

Taking into consideration a temporary time change in frequency of the signal and making a comparison with the existing database of various scenarios of gravitational waves emission, we made a conclusion that this signal corresponds to the confluence of two black holes, hence, we could identify the masses of the black holes. The expression for the dependence of the observed signal frequency on time, obtained in Newtonian approximation of the general theory of relativity [4] (see also [5]), was the main expression used to determine the mass of the collapsing black holes:

$$\left(\frac{(m_1 m_2)^3}{m_1+m_2}\right)^{\frac{1}{5}} = \frac{c^3}{G}\left(\frac{5}{96}\pi^{-\frac{8}{3}} f^{-\frac{11}{3}} \frac{df}{dt}\right)^{\frac{3}{5}}, \tag{1}$$

where m_1, m_2 – are the masses of confluent black holes, G – is the gravitational constant, c – is the speed of light, $f(t)$ – is the rotation frequency of the masses of black holes around the common center of inertia. Solving the equation (1) with respect to frequency, we obtain

$$\frac{96\pi^{8/3}}{5}\frac{GM}{c^3}(t_2-t_1) = \frac{3}{8}\left(f^{-8/3}(t_1) - f^{-8/3}(t_2)\right), \tag{2}$$

where t_2, t_1 – are the points of time at which the frequency value $f(t_2)$, $f(t_1)$ is measured, respectively $M \equiv \left(\frac{(m_1 m_2)^3}{m_1+m_2}\right)^{\frac{1}{5}}$. Using the obtained relation (2) in [1], the most probable values of the masses of black holes were found according to the obtained frequency values $f(t_2)$, $f(t_1)$. They appeared to be equivalent to $36^{+5}_{-4}M_\odot$ and $29^{+4}_{-4}M_\odot$,where $M_\odot$ is the mass of the sun.

In this work we propose a slightly different method for determining the most probable blackhole mass values, and we need to know only one, the lowest, frequency of the observed gravitational waves and the time lag between the point of measuring the lowest frequency and the point of the double black hole collapse. In this case, we do not need to know the highest frequency of gravitational radiation. The fact is that the maximum frequency value occurs immediately before the collapse when the velocity values of the confluent black holes are comparable with the speed of light (for example, in the case of events recorded in [1], this value reaches 0.4-0.5 the speed of light) and as it is shown in [5], relativistic corrections become significant. Knowing the lower frequency corresponds exactly to Newtonian approximation used in deriving the ratio (2). Certainly, the time interval in a strong gravitational field of a black hole is known [4] to be changing, however, time is the integral value and the longer the time lag in which the observations are carried out is, the smaller is the proportion of time changes. In contrast, the oscillation frequency in a gravitational wave is an instantaneous parameter and its value in the strong gravitational field depends strongly on the gravitational field value [4,5].

We now obtain a slightly different expression for determining the masses of collapsing black holes. We consider the motions in Newtonian approximation, when we meet the conditions of black hole slowness compared to the speed of light, i.e. the condition $GM/r \ll c^2$ is fulfilled. Then, using the relationship between the rotation frequency $\omega(t)$ of the two masses around the common center of inertia

$$\omega^2(t)\, r^3(t) = G(m_1 + m_2), \tag{3}$$

where $r(t)-$ is the radius of the orbit, as well as the expression for the binding energy of the two masses $E(t) = -G\dfrac{m_1 m_2}{2r(t)}$ and the expression for the energy loss due to the emission of gravitational waves

$$\frac{dE(t)}{dt} = -\frac{32\,G}{5c^5}\left(\frac{m_1 m_2}{m_1 + m_2}\right)^2 r^4(t)\,\omega^6(t) \tag{4}$$

we obtain an equation for the time change in the rotation radius [3]

$$\frac{dr(t)}{dt} = -\frac{64\,G^3\, m_1\, m_2 (m_1 + m_2)}{5c^5} r^{-3}(t). \tag{5}$$

Equation (5) allows us to find the time lag from the start of the measurement t_1 until the moment when the collapse occurred t_2 :

$$\Delta t = \frac{5c^5}{256\, G^{5/3} m_1\, m_2\, (m_1 + m_2)}\left(r^4(t_1) - r^4(t_2)\right), \qquad \Delta t \equiv t_2 - t_1. \tag{6}$$

Equation (6) and formula (2) are very similar, if in (2) we drop member $f(t_2)$, and in (6) we drop member $r^4(t_2)$, they coincide. The latter is the direct consequence of Newtonian approximation, when approaching the collapse $r \to 0$, $f \to \infty$, but near the collapse the speeds of black holes are strongly relativistic and Newtonian approximation is no longer valid. However, in the time lag (6), the proportion of non-Newtonian relativistic time lag becomes smaller, as the time lag of the observation is longer. For point masses the collapse means that the distance between the rotating mass tends to zero and, consequently, the frequency f tends to infinity (in non-relativistic approximation, see formula (3)). After the two masses collapse and the value of the quadrupole moment of the system decreases significantly, as it is clear from the observations [1], the intensity of gravitational waves emission falls sharply, which is quite clear from the general physical considerations. This circumstance makes it possible to more accurately determine the point of time when the confluence of two black holes occurred.

The given graph of gravitational waves frequency and its changes depending on the time clearly shows that this point of time corresponds to the change of the derivative sign $\dfrac{df}{dt}$ or, more precisely, to the point of time t_2 at which the second derivative is equal to zero. Thus, the observations give an opportunity to more accurately determine the time when there was collapse and thus more accurately determine the interval $\Delta t = t_2 - t_1$. According to the observations presented in [1], we conclude that the interval is approximately $\Delta t \simeq 0{,}1754\,\text{sec}$, and considering that the rotation frequency of the two black holes around the total moment of inertia is equal to one half of the frequency of gravitational wave emission, i.e. 17.5 Hz, for the mass values $36^{+5}_{-4}\ M_{\odot}$ and $29^{+4}_{-4}\ M_{\odot}$ given in [1], we obtain the value for the interval $\Delta t \simeq 0{,}1754 сек$ from the formula (6). Thus, there is a good agreement between observations and estimates for the masses of collapsing black holes given in [1].

Given that the value of the rotation radius at the time of the collapse approximates to the value of Schwarzschild radius for the total mass of the black holes, taking into account the weight loss in the emission of gravitational waves, we can assume that $r\ (t_1) \gg r\ (t_2)$, and then the formula (6) is even simpler:

$$\Delta t \simeq \frac{5c^5 (m_1 + m_2)^{1/3}}{256\, G^{5/3} m_1\, m_2\, \omega_{\min}{}^{8/3}(t_1)}, \tag{7}$$

where $\omega_{\min}\ (t_1)$ is the minimum value of the rotation frequency of black holes, which is equal to the half of the frequency of the observed gravitational waves. This simple formula is more convenient for selecting the values of the masses of the collapsing black holes.

As we continue to improve the sensitivity of LIGO type gravitational antennas, we need to express our view concerning the use of mirrors in Fabry-Perot cavity. We aim at improving the reflectance coefficient of laser light mirrors in Fabry-Perot cavity and we result in a significant increase in circulating laser power, which now reaches the values of 100 kW, and in the future is likely to increase even more - up to 830 kW. It is clear that even with a small absorption in the reflective multilayer structures, there is not only the problem of heat dissipation, but also the problem of misalignment of the cavity itself due to nonuniform mirror heating, and therefore it is necessary to introduce the additional heat in order to prevent misalignment of the system.

Taking this into account it is necessary to consider other approaches to creating highly reflective mirrors that can withstand high power laser radiation. In this regard, we should pay attention and explore the possibility of obtaining highly reflective extended structures by means of creating the refractive index in the medium (i. e. correct periodic) of sinusoidal change structure. The amplitude of the change in refractive index is small, there are no sharp jumps, and high reflectance is achieved through the use of an extended periodic structure (sinusoidal Bragg mirror).

Such structures were studied in detail previously (see [6-8]), when in the mediumthere is a propagating sound wave and due to the effect of photoelasticity, there occurs periodic variation of the refractive index, on which the incident laser light diffracts "back". Certainly, the creation of such a periodic structure by means of a sound wave is unlikely to be suitable for LIGO type interferometers. However, the use of heterogeneous media, or one-dimensional photonic crystals, in which the use of a particular technology provides the necessary periodic structure of changes in the dielectric permeability of the medium, allows us [9] to solve many problems in the mirror interferometers such as LIGO. Let us briefly consider the "back" diffraction phenomenon of the of the incident laser light on the mediaum, when the dielectric permiability of the medium $\varepsilon(x)$ varies according to the law:

$$\varepsilon(x) = \varepsilon_0 + \Delta\varepsilon(x), \qquad \Delta\varepsilon(x) = \Delta\varepsilon\, \mathrm{Cos}(qx), \quad \Delta\varepsilon(x) \ll \varepsilon_0. \tag{8}$$

Here $\Delta\varepsilon$ is the "amplitude" of changes in the dielectric permeability of the medium, q – is the wave vector of the periodic structure of the medium. Maxwell's equations for the incident and reflected waves propagating in the opposite direction, in the approximation of slowly varying amplitudes

$$E^{+}(x,t)=\frac{1}{2}E_{0}^{+}(x)Exp(i\omega t-ikx)+k.c.\ E^{-}(x,t)=\frac{1}{2}E_{0}^{-}(x)Exp(i\omega t+ikx)+k.c.$$

are written in the form:

$$\frac{dE^{+}(x)}{dx}=i\,k\Delta\varepsilon\,E^{-}(x)Exp(i\Delta k\,x),\qquad \frac{dE^{-}(x)}{dx}=-i\,k\Delta\varepsilon\,E^{+}(x)Exp(-i\Delta k\,x), \quad (9)$$

ω, k – are frequency and wave vector of the laser radiation propagating within Fabry-Perot cavity, $\Delta k=2k-q$ – is the departure from synchronism conditions. Solving the equation (9) with the boundary conditions: $E^{+}(0)=E_{0}$, $E^{-}(L)=0$ where L is the size of the structure, for the amplitude of the reflected wave propagating in the opposite direction with respect to the direction of the incident wave, we obtain the following expression:

$$E^{-}(\Delta k)=\frac{2i\,k_{0}\Delta\varepsilon E_{0}}{\Delta k-i\sqrt{-\Delta k^{2}+4k_{0}^{2}\Delta\varepsilon^{2}}\ \mathrm{Coth}\left(\frac{1}{2}L\sqrt{-\Delta k^{2}+4k_{0}^{2}\Delta\varepsilon^{2}}\right)} \quad (10)$$

The amplitude and phase dependance of the reflected wave on the extent of the departure from synchronism condition in accordance with the formula (1) is shown in Fig. 2. It also shows numerical values which were used when making the graphs. We see that at relatively low values of refractive index changes at a sufficient length of the structure we have a reflection coefficient which is very close to the unity. Withvalues indicated in Fig. 2, the amplitude of the reflected wave at $\Delta k=0$ will be $E^{-}=0.9999930E_{0}$ (without radiation attenuation in the medium). Formula (10) is easily generalized in the case of an absorbing medium. Thus, if we use periodic structures in such a way, when radiation occupies almost the entire volume of the mirror, the heat is more easily removed. It should be noted that formula (10) was obtained under the condition of a spatially homogeneous distribution of the sinusoidal periodic structure, but there exist more interesting practical approaches when the structure itself is still modulated (see[7]). At a certain modulation type (see [7] as an example), there occur exponentially narrow radiation transmission lines with controlled intensity, which can be very useful for the semireflecting Fabry-Perot mirror, as well as in the problem of the laser mode purification. It is necessary to point out that if Fabry-Perot cavity has such mirrors with a very narrow bandwidth line, it allows us to create new optical schemes for direct detecting of the gravitational waves.

With regard to the technology of periodic structures, we can point out the following possibilities. a) First, we create a periodic structure consisting of alternating layers (sheets) of a transparent insulator such as fused quartz, and in its sheets there were sequentially and alternatively inserted nano-particles of impurities (metals or insulators with the large dielectric permeability value), changing the dielectric permeability of the sheet. Then, we heat the stack of the sheets to the desired temperature and pass this workpiece through a series of cylindrical waves, reducing its cross-sectional dimension to the desired value. Such technology is already used to create a gradient optical fiber. b) There is another approach, based on the fact that the heated heterogeneous medium containing the necessary quantity of nano particles is placed in a microwave resonator of electromagnetic oscillations, in which nano-particles affected bythe electric field of a standing electromagnetic wave, are redistributed in space in accordance with the structure of the field, and then such a workpiece passes through a series of waves changing its cross-sectional dimensions to the desired value. There are other methods of obtaining periodic structures associated with exposure of high-power laser or acoustic waves on heterogeneous media.

Concerning the problem of hanging the mirrors the following should be noted. It is known that the sufficiently powerful optical radiation emitted on the free-suspended mirror, polarizes it and the mirror appears to have dipole moment. This polarization occurs when a mirror has a number of external charges [10], which may result from many causes: space radiation affects the mirror itself, as well as the surrounding structures, residual gas molecules hit the surface of the mirror, and other reasons. Occurrence of polarization in an insulator, a semiconductor or even in some metals having oxidized film on the surface has been regularly observed experimentally [11].

The appearance of dipole moment in a mirror should lead to a further link with the surrounding equipment (dipole - dipole interaction), especially with conductive metal structures. This relationship can be very undesirable, particularly in a laser beam modulation. This is a new effect, which unlike the usual Coulomb charge interaction on the mirror, interacts with metal structures, when there is laser radiation.

The effect of appearance of dipole moment in the mirror, or in other words, its polarization, apparently, has not been observed until now. We believe this effect may be of significant importance for the problem of alignment of the optical interferometer scheme and it is not limited to recording the light pressure on the mirror. In this case we are talking about the effect of volume ponderomotive forces from the laser emissionon the mirror itself, and its multilayered periodic structure.

It is well known (see. [4], §15, and [10], §§32-34), that an insulator is acted upon the volume force from the electric field.

$$\mathbf{f}(x)=\frac{1}{8\pi}\nabla\left(<\mathbf{E}^2(x,t)>\rho\frac{\partial\varepsilon}{\partial\rho}\right)-\frac{<\mathbf{E}^2(x,t)>}{8\pi}\nabla\varepsilon+<q\,\mathbf{E}(x,t)>, \qquad (11)$$

here $\mathbf{f}(x)$ is the volume force density acting on the medium, $<\mathbf{E}^2(x,t)>$ is the time average of the square of the laser radiation amplitude, $\frac{\partial\varepsilon}{\partial\rho}$ is the density derivative of the dielectric permeability of the medium (the medium is assumed to be isotropic), q is the volume density of external charges induced in the mirror. Expression (11), unlike the equations for ponderomotive forces given in the monographs [3,11], includes the first two time averaged terms, because it is believed that there are no free light particles (electrons), which could follow the wave field oscillations. As for the last term in (11), ifthe mirror has free charges, it is necessary to carry out its averaging considering the fact that the charge density $q=q(x,t)$ under the action of the wave field can be redistributed (see references [12-14]). The force (11) is of the greatest problem for the time modulated laser signal, in this case different resonances are quite possible, and they need to be given special attention.

It is known [10,12], in addition to the volume force, there is also a surface force which can be obtained from the formula (11) by integrating according to the volume of the mirror. Then we can show that there is a surface density of the force acting on the boundary of two media with different dielectric permeability values in each of the layers. This surface force density "puts pressure" on the boundary, and this pressure is determined by the following:

$$f(x)=\frac{1}{2}<E(x,t)^2>\left(\varepsilon_1-\varepsilon_2\right), \qquad (12)$$

where $\varepsilon_1, \varepsilon_2$ are the values of dielectric permeability of the adjacent layers of the periodic structure. With the spread of the radiation through the periodic structure, the value of the tangential field component is reduced, and the pressure sign, as it can be seen from the formula (12) varies along the structure, creating different voltages. Such spatial variation of the pressure can lead to a number of effects: changes in the thickness of the layers and, as a consequence, deviation from the synchronism conditions, or the destruction of the periodic structure itself, and if there is temporal modulation of laser emission intensity, it can result in appearance of acoustic vibrations. (Note, that such an obvious mechanism of high-frequency acoustic vibrations excitation in a medium with a periodic change of dielectric permeability by the modulated laser radiation, apparently, has never been examined before. It is clear that this problem has to be studied separately).

Thus, at the boundary between two media with different dielectric permeability there appears a force, leading to the deformation of these layers, and in the presence of the laser intensity modulation, there can be high-frequency acoustic waves. All these effects need to be considered when taking measurements.

4. Concerning the question, whether our work [2] is known to professionals working on creation of resonant antennas for the reception of gravitational waves, and making the first attempts to detect gravitational waves (J.Weber, 1960), it is necessary to note the following.

In 1963, my teacher VL Ginzburg participated in a conference on gravitation in Poland (Warsaw) [15]. There he gave a talk in which he criticized the resonance detectors of gravitational waves relying on our work. It is known that the resonance gravitational wave detectors are most sensitive at the well-defined resonance frequency, and have a very narrow spectral band. The frequency of possible sources of gravitational waves is unknown, moreover, the direction from which gravitational waves can arise or their polarization is unknown as well. It is these features of resonance antennas, in our opinion, that led to the failure of detecting gravitational waves by such a method, and we noted this in our work [2], in which we also proposed to use broadband laser interferometers.

This conference was attended by Joseph Weber and in August 1963 he published a report [16], which contained a response to this "criticism" in our work [2]. J. Weber referred to V.L. Ginzburg'swork [15] and our work [2]. Unfortunately, Joseph Weber' paper [16] is not cited by many researchers, but Felix Pirani' report [17] was apparentlycited for the first time by Joseph Weber in his work [16], and it is sometimes referred to as the work relating to the problem of the direct detection of gravitational waves.

Joseph Weber's pioneer experiment presented the antenna as an aluminum cylinder 2 m long and 1 m in diameter, suspended on steel wires; resonance frequency of the antenna was 1660 Hz, the amplitude sensitivity of piezoelectric detectors was about10^{-14} sm. J. Weber used two detectors working in coincidence, and he reported the detection of a signal, whose source was most likely the center of the Galaxy. However, no independent experiments confirmed Weber's observations. Still it should be noted that J. Weber's pioneering work played an important role in developing and understanding the problem of direct detection of gravitational waves, and attracted much attention from researchers. Among detectors currently operating on this principle we can mention the spherical antenna MiniGRAIL (Leiden University, The Netherlands), the antenna ALLEGRO (Baton Rouge, Louisiana, USA), with which Joseph Weber worked he past, AURIGA (Rome, Italy), the receiver temperature in this antenna is $0{,}1^{o}K$, NAUTILUS (Rome, Italy), the EXPLORER (Switzerland) and Agra (Russia).

Journal of Physics: Conference Series **731** (2016) 012011 doi:10.1088/1742-6596/731/1/012011

References

[1] Abbott B P 2016 et al. *Phys. Rev. Lett.* **116** 061102
[2] Gertsenshtein M E, Pustovoit V I 1962 *Journal of Theoretical and Experimental Physics* **43** pp 605-607
[3] Weiss R 1972 Electromagnetically coupled broadband gravitational antenna. *Quarterly Report of the Research Laboratory for Electronics, MIT Report* **105**
[4] Landau L D, Lifshits E M 1967 *Field theory* (Moscow: Nauka Publ.)
[5]Blanchet L, Damour T, Iyer B R, Will C M and Wiseman A G 1995 *Phys. Rev. Lett.* **74** 3515
[6] Pustovoit V I Special case of light collinear diffraction on sound waves in crystals. *Optical memory and Neural Networks (Information Optics)* **4** no. 2
[7] Afanasev A M, Guliaev Iu V, Pustovoit V I 2003 *Radiotekhnika I elektronika* **49** pp 1526-1531
[8] Pustovoit V I 2014 *Selected works* (Moscow: Nauka Publ.) pp 42-53
[9] Pustovoit V I 2006 *Proceedings of RAS* **407** no. 4 pp 472-477
[10] Pustovoit V I, Borisov M, Ivanov O 1989 *Phys.Lett.A* **135** no.11
[11] Tamm I E 2003 *Basics of the theory of electricity* (Moscow: Fizmatlit Publ.)
[12] Solunin S A, Solunin A M, Solunin M A 2009 *JETP Letters* **35** no.14 pp 48-53
[13] Aleksandrov A F, Bogdankevich L S, Rukhadze A A 1988 *Basics of plasma electrodynamics* (Moscow: High School Publ.) p 424
[14] Bolotovskii B M, Serov A V 1994 *Advances in Physical Sciences* **164** no. 5 pp 545-549
[15] Ginsburg V L 1963 *Inter. Conf. on Gravitation Warsaw July*
[16] Weber J 1963 Remarks on gravitational experiments. *IL Nuovo Cimento* **29** no. 4 pp 930-934
[17] Pirani F A 1956 *Acta Physica Polonica* no. 15 pp 389–405

Solitons in Skyrme – Faddeev spinor model and quantum mechanics

Y Rybakov
Dept. Theor. Phys. Mech., Peoples' Friendship University of Russia, Moscow, Russia
E-mail: soliton4@mail.ru

Abstract. We discuss the possibility of unification of Skyrme and Faddeev approaches for the description of baryons and leptons respectively as topological solitons within the scope of 16-spinor model. The motivation for such a unification is based on a special 8-semispinor identity invented by the Italian geometrician F. Brioschi. This remarkable identity permits one to realize baryon or lepton states through the effect of spontaneous symmetry breaking emerging due to special structure of the Higgs potential in the model. At large distances from the particle – soliton small excitation of the vacuum satisfies Klein – Gordon equation with some mass that permits one to establish the correspondence with quantum mechanics in special stochastic representation of the wave function for extended particles – solitons. Finally, we illustrate the peculiar properties of stochastic representation by the famous T. Young's experiment with n slits in soliton realization.

1. Introduction. Brioschi identity and 16-spinor field model

The Skyrme's fruitful idea [1] to describe baryons as topological solitons was based on the identification of the baryon number $\mathbb{B}$ with the topological charge of the degree type $\mathbb{B} = \deg(S^3 \to S^3)$, which serves as the generator of the homotopy group $\pi_3(S^3) = \mathbb{Z}$. The similar idea to describe leptons as topological solitons was announced by Faddeev [2], who identified the lepton number $\mathbb{L}$ with the Hopf invariant Q_H, which serves as the generator of the homotopy group $\pi_3(S^2) = \mathbb{Z}$. The unification of these two approaches was suggested in [3], baryons and leptons being considered as two possible phases of the effective 8-spinor field model, for which the special 8-spinor Brioschi identity [4] holds:

$$j_\mu j^\mu - \tilde{j}_\mu \tilde{j}^\mu = s^2 + p^2 + \vec{v}^2 + \vec{a}^2, \tag{1}$$

where the quadratic spinor quantities are introduced:

$$s = \overline{\Psi}\Psi,\quad p = i\overline{\Psi}\gamma_5\Psi,\quad \vec{v} = \overline{\Psi}\vec{\lambda}\Psi,\quad \vec{a} = i\overline{\Psi}\gamma_5\vec{\lambda}\Psi,\quad j_\mu = \overline{\Psi}\gamma_\mu\Psi,\quad \tilde{j}_\mu = \overline{\Psi}\gamma_\mu\gamma_5\Psi,$$

with $\overline{\Psi} = \Psi^+\gamma_0$ and $\vec{\lambda}$ standing for Pauli matrices in the isotopic spinor space. Here $\gamma_\mu,\ \mu = 0,1,2,3,$ designate the unitary Dirac matrices acting on Minkowski spinor indices. One can verify the time-like character of the Dirac 4-current j_μ and, in view of the identity (1), use the special structure of the Higgs potential V implying the spontaneous symmetry breaking in our model:

$$V = \frac{\sigma^2}{8}(j^\mu j_\mu - æ_0^2)^2, \tag{2}$$

where σ and $æ^0$ stand for some constant parameters. If one searches for localized soliton-like configurations, one should use the natural boundary condition at space infinity:

$$\lim_{|\vec{x}|\to\infty} j_\mu j^\mu = æ_0^2, \tag{3}$$

which determines the fixed (vacuum) point in the phase space of our model. In particular, if for the vacuum state Ψ_0 one gets $s(\Psi_0)\neq 0$, then the chiral invariant $s^2+\vec{a}^2$ determines sphere S^3 as the field manifold, that corresponds to the Skyrme model and the baryon phase. If $v_3(\Psi_0)\neq 0$, then the $SO(3)$ invariant $\vec{v}^2$ determines the sphere S^2 as the field manifold, that corresponds to the Faddeev model and the lepton phase.

However, the vacuum state Ψ_0 should be universal, that is common both for baryon and lepton phases. It means that conditions $v_3(\Psi_0)\neq 0$ and $s(\Psi_0)\neq 0$ should be compatible. Taking into account that the Brioschi identity (1) holds only for 8-spinor fields, one concludes that the realistic model should include the two 8-spinor fields Ψ_1 and Ψ_2, which can be naturally unified forming 16-spinor field $\Psi=\Psi_1\oplus\Psi_2$. Finally, each topological sector should be described by the effective 8-spinor field.

In view of these arguments, using the analogy with Skyrme – Faddeev model, we consider the following Lagrangian density for the 16-spinor field model resembling the one used in [3]:

$$L_{spin}=\frac{1}{2\lambda^2}\overline{\partial_\mu\Psi}\gamma^\nu j_\nu\partial^\mu\Psi+\frac{\varepsilon^2}{4}f_{\mu\nu}f^{\mu\nu}-V\,, \tag{4}$$

where $f_{\mu\nu}$ stands for the anti-symmetric tensor of Skyrme – Faddeev type:

$$f_{\mu\nu}=(\overline{\Psi}\gamma^\alpha\partial_{[\mu}\Psi)(\overline{\partial_{\nu]}\Psi}\gamma_\alpha\Psi), \tag{5}$$

with λ and ε being constant parameters of the model. It should be underlined that the first term in (4) is similar to the sigma-model one and includes the projector $P=\gamma^0\gamma^\nu j_\nu$ on the positive energy states. We also remark that the Dirac 16-spinor current conserves its time-like character due to additive property $j_\mu=j_\mu^{(1)}+j_\mu^{(2)}$.

In 16-spinor space we introduce the two kinds of internal Pauli matrices:

$$\vec{\lambda}=I_4\otimes\vec{\sigma}\otimes I_2,\quad \vec{\Lambda}=I_8\otimes\vec{\sigma}, \tag{6}$$

where I_n designates the n-dimensional unity and $\vec{\sigma}$ - usual Pauli matrices. We suppose also that the vacuum state has the following structure:

$$\Psi_0=\Psi_{10}\oplus 0, \tag{7}$$

with the projective property:

$$\Lambda\Psi_0=0,\quad \Lambda=\frac{1}{2}(1-\Lambda_3). \tag{8}$$

Now the main question arises: how to choose the lepton sector or the baryon one? To find the answer, we consider the topological structure of our model implying the distinction between these sectors. First, we pay attention to the fact that in the lepton sector $\vec{a}=i\overline{\Psi}\gamma_5\vec{\lambda}\Psi=0$ that implies B = 0. This condition can be satisfied if we suppose the mirror symmetry of the lepton sector:

$$\Psi\to\gamma_0\Psi. \tag{9}$$

Therefore, if we introduce the following structure of the 16-spinor:

$$\Psi=\bigoplus_{j=1}^{2}(\varphi_j\oplus\chi_j\oplus\xi_j\oplus\zeta_j), \tag{10}$$

where $\varphi_j, \chi_j, \xi_j, \zeta_j$ stand for 2-spinors, then (9) implies the invariance condition $\Psi = \gamma_0 \Psi$, that is $\varphi_j = \chi_j$, $\xi_j = \zeta_j$ for the Weyl representation of γ-matrices. Thus, we conclude that the lepton sector is described by the effective 8-spinor:

$$\Psi_L = \bigoplus_{j=1}^{2} (\varphi_j \oplus \varphi_j \oplus \xi_j \oplus \xi_j), \tag{11}$$

which is evidently consistent with the Brioschi identity, since the new vector $\vec{V} = \overline{\Psi}\vec{\Lambda}\Psi \neq 0$.

In particular, one gets

$$j_0 = 2\sum_{j=1}^{2} (|\varphi_j|^2 + |\xi_j|^2); \quad \vec{j} = 0; \quad V_1 = 4\,\mathrm{Re}\,(\varphi_1^+ \varphi_2 + \xi_1^+ \xi_2);$$
$$V_2 = 4\,\mathrm{Im}(\varphi_1^+ \varphi_2 + \xi_1^+ \xi_2); \quad V_3 = 2(|\varphi_1|^2 + |\xi_1|^2 - |\varphi_2|^2 - |\xi_2|^2),$$

with the evident property:

$$\frac{1}{4}(j_0^2 - \vec{V}^2) = 4(|\varphi_2|^2 + |\xi_2|^2)(|\varphi_1|^2 + |\xi_1|^2) - 4|\varphi_1^+ \varphi_2 + \xi_1^+ \xi_2|^2 \geq 0$$

due to the Schwartz inequality. In view of the boundary condition (3) we conclude that the correspondent localized configuration can be endowed with the nontrivial Hopf index $Q_H = \mathbb{L}$, the S^2-manifold being determined by the invariant structure $\vec{V}^2$.

Let us now consider the baryon sector. In this case we take into account the charge independence of strong interactions, which is equivalent to the mirror symmetry in the isotopic space:

$$\Psi \to \gamma_0 \gamma_5 \gamma_2 \lambda_2 \Psi^*. \tag{12}$$

Inserting (10) into the invariance condition for the group (12), one finds the following structure for the baryon sector 16-spinor field:

$$\Psi_B = \bigoplus_{j=1}^{2} (\varphi_j \oplus \chi_j \oplus i\sigma_2 \varphi_j^* \oplus i\sigma_2 \chi_j^*), \tag{13}$$

which is again equivalent to the effective 8-spinor. In this case we find the following bilinear spinor quantities determining the topology of the baryon sector:

$$j_0 = 2\sum_{j=1}^{2} (|\varphi_j|^2 + |\chi_j|^2); \quad \vec{j} = 0; \quad s = 4\sum_{j=1}^{2} \mathrm{Re}\,(\varphi_j^+ \chi_j);$$
$$a_1 = -4\sum_{j=1}^{2} \mathrm{Re}\,(\varphi_j^T \sigma_2 \chi_j); \quad a_2 = -4\sum_{j=1}^{2} \mathrm{Im}\,(\varphi_j^T \sigma_2 \chi_j); \quad a_3 = 4\sum_{j=1}^{2} \mathrm{Im}\,(\varphi_j^+ \chi_j).$$

Therefore, we can calculate the chiral invariant structure determining S^3-manifold:

$$s^2 + \vec{a}^2 = 16\,(|\varphi_1^+ \chi_1 + \varphi_2^+ \chi_2|^2 + |\varphi_1^T \sigma_2 \chi_1 + \varphi_2^T \sigma_2 \chi_2|^2).$$

Using the Schwartz inequality, one can prove that $j_0^2 \geq s^2 + \vec{a}^2$. This structure is again consistent with the Higgs potential (2).

2. Interaction with physical vector fields

The Lagrangian (4) admits very large group of transformations of the 16-spinor field Ψ, including phase transformations with some charge generator Γ_e and left, right isotopic rotations with generators $P_{L,R}\lambda^a \Lambda/2$; $a = 1,2,3$, where $P_{L,R} = (1 \pm \gamma_5)/2$ stand for the corresponding projectors. These symmetries give rise to the interactions with the electromagnetic field A_μ

and left, right Yang – Mills fields $A_\mu^{aL,R}$ respectively, the latter ones being responsible for the strong interactions. Due to the general gauge invariance principle these interactions can be included in the Lagrangian through the extension of the derivative $\partial_\mu \to D_\mu$, where

$$D_\mu \Psi = \partial_\mu \Psi - i e_0 \Gamma_e A_\mu \Psi + (A_\mu^L + A_\mu^R)\Psi. \tag{14}$$

We adopt the following structure of Γ_e and $A_\mu^{aL,R}$:

$$\Gamma_e = P_3 \Lambda, \quad P_3 = (1-\lambda_3)/2, \quad A_\mu^{L,R} = P_{L,R} \frac{e_{1L,R}}{2i} A_\mu^{aL,R} \lambda^a \Lambda, \tag{15}$$

which is in accordance with the breaking of isotopic symmetry by the electromagnetic interaction, e_0, e_{1L}, e_{1R} being the corresponding coupling constants. However, the fields $A_\mu^{aL,R}$ seem to be not independent due to the natural condition that Yang – Mills fields should not give any contribution to the lepton sector. This condition can be satisfied, in view of (9) and (14), only if the sum $A_\mu^L + A_\mu^R$ is proportional to γ_5. Taking into account (15), one concludes that this is possible, only if the following constraint holds:

$$e_{1L} A_\mu^{aL} + e_{1R} A_\mu^{aR} = 0. \tag{16}$$

In view of (16) one can easily find that

$$A_\mu^L + A_\mu^R = \frac{e_{1L}}{2i} A_\mu^{aL} \gamma_5 \lambda^a \Lambda \equiv g_0 \gamma_5 \mathrm{A}_\mu, \tag{17}$$

where the denotation is used:

$$e_{1L} \equiv g_0; \qquad \mathrm{A}_\mu \equiv \frac{\lambda^a}{2i} \Lambda A_\mu^a; \quad A_\mu^{aL} \equiv A_\mu^a.$$

If one introduces the intensity of the Yang – Mills fields:

$$F_{\mu\nu}^{L,R} = \partial_\mu A_\nu^{L,R} - \partial_\nu A_\mu^{L,R} + [A_\mu^{L,R}, A_\nu^{L,R}]$$

and use the natural supposition that $e_{1L} = e_{1R}$, then the standard Yang – Mills Lagrangian can be rewritten, in view of the constraint (16), as follows:

$$L_{YM} = \frac{1}{32\pi g_0^2} Sp(F_{\mu\nu}^L F^{\mu\nu L} + F_{\mu\nu}^R F^{\mu\nu R}) = \frac{1}{32\pi} Sp\{(\partial_\mu \mathrm{A}_\nu - \partial_\nu \mathrm{A}_\mu)(\partial^\mu \mathrm{A}^\nu - \partial^\nu \mathrm{A}^\mu) +$$
$$+ g_0^2 [\mathrm{A}_\mu, \mathrm{A}_\nu][\mathrm{A}^\mu, \mathrm{A}^\nu]\},$$

where the symbol *Sp* signifies the corresponding matrix trace.

Now the extended derivative (14) takes the form:

$$D_\mu \Psi = \partial_\mu \Psi - i e_0 \Gamma_e A_\mu \Psi + g_0 \gamma_5 \mathrm{A}_\mu \Psi, \tag{18}$$

the effective Yang – Mills interaction being pseudo-vector one.

3. Solitons and stochastic representation of quantum mechanics

Let us consider small excitations of our soliton near the vacuum: $\Psi = \Psi_0 + \xi$, $\xi \to 0$ as $|\vec{x}| \to \infty$. Then the linearized equation for ξ after the substitution $\xi = k\,\Psi_0$ takes the form:

$$\partial_\mu \partial^\mu k + \frac{1}{2} M_0^2 (k^* + k) = 0, \quad M_0 \equiv 2\sigma\lambda\, æ_0. \tag{19}$$

Thus, from (19) one derives the following equations for real and imaginary parts of $k = k_1 + i k_2$:

$$\partial_\mu \partial^\mu k_1 + M_0^2 k_1 = 0, \quad \partial_\mu \partial^\mu k_2 = 0. \tag{20}$$

According to (20), our model admits two types of vacuum excitations: massive and massless ones. For massive soliton the first equation in (20) coinciding with the well-known Klein – Gordon one, the spinor excitation ξ could be interpreted as the wave function of the point-like particle representing the motion of the center of the soliton, if the mass parameter M_0 corresponded to the real mass M of the particle - soliton (or inverse Compton length in natural units $\hbar = c = 1$). To satisfy this condition, we first introduce the interaction with the gravitational field via the new extended derivative generalizing (14):

$$\nabla_\mu \Psi = (D_\mu - \Gamma_\mu)\Psi, \tag{21}$$

where Γ_μ stands for spinor connection with gravity. Then we intend to generalize the Higgs potential (2), in which the constant multiplier σ^2 will be replaced with the special invariant:

$$\sigma^2 = -\frac{16D^3}{\lambda^2 G^2 K^2} æ_0^{-2}, \tag{22}$$

where G stands for the Newton gravitational constant, D is thesigma-model part of the Lagrangian L_{spin}:

$$D = \overline{\nabla_\mu \Psi} \gamma^\alpha j_\alpha \nabla^\mu \Psi \tag{23}$$

and K is the so-called Kretschmann invariant constructed with the help of the Riemannian curvature tensor:

$$K = \frac{1}{48} R_{\mu\nu\sigma\lambda} R^{\mu\nu\sigma\lambda}. \tag{24}$$

The validity of the choice (22) can be verified by calculating invariants (23) and (24) for Schwarzschild metric at large distances $r = |\vec{x}| \to \infty$:

$$D = -\frac{r_g^2}{4r^4} æ_0^2; \quad K = \frac{r_g^2}{r^6}; \quad r_g = GM. \tag{25}$$

The resulting Lagrangian of our 16-spinor model reads:

$$L = L_{spin} + L_{em} + L_{YM} + L_g, \tag{26}$$

where L_{em} stands for the Maxwellian electromagnetic part and the gravitational Lagrangian L_g coincides with the Einstein one: $L_g = R/(16\pi G), R$ being the scalar curvature.

It is worth-while to stress that the gravitational field plays an important role in our model, since the wave-particle duality principle of quantum mechanics has the gravitational origin. As for the vacuum excitation ξ, one can prove [5, 6] that it plays the role of the wave function in the special stochastic representation of quantum mechanics. To this end, we discuss the fundamental problem of statistical description of solitons stimulated by experimental observations confirming wave - particle duality relations and other quantum properties. The usefulness of the stochastic representation of the wave function will be illustrated by the famous T. Young experiment in soliton realization.

Let us consider an impenetrable screen with n slits, for which the width $d >> l$, where $l^3 = V_0$ denotes the proper volume of particles - solitons pushed to the screen one by one with some initial velocity. Our aim is to find the probability $dW(x) = w(x)dx$ for observing a center of a particle in the vicinity dx of some point x behind the screen at large distance from it. Let us also introduce the probability $dW_j(x) = w_j(x)\, dx$ for the soliton to pass through the j-th slit, and therefore $dW = \sum_{j=1}^{n} dW_j$. However, the wave - particle duality means that there should be the influence of the k-th slit on the j-th one, i. e. on the passing through the j-th slit. That is why the following relation holds:

$$w = \sum_{j=1}^{n} w_j = \sum_{j,k=1}^{n} w_{jk} \,. \tag{27}$$

As well known, for classical particles $w_{jk} = 0$, if $j \neq k$, contrary to quantum particles.

How to construct the wave function for 1-particle states in stochastic representation? First we introduce the Lagrangian density $L(\phi, \partial_\mu \phi)$ for an abstract real field $\phi(t, \vec{x})$ with an arbitrary number of components. Suppose that the equations of motion admit 1-particle soliton-like solution $\phi(t, \vec{x})$, define its canonical momentum $\pi(t, \vec{x}) = \partial L / \partial \dot{\phi}$ and also the auxiliary complex function

$$\varphi(t, \vec{x}) = \frac{1}{\sqrt{2}} (\nu \phi + i \pi / \nu)\,, \tag{28}$$

where the number ν can be found from the following normalization condition:

$$\hbar = \int d^3x \, |\varphi|^2 , \tag{29}$$

with $\hbar$ being the Planck constant. Assuming the experiment with scattered soliton to be repeated $N >> 1$ times, we define the wave function $\psi(t, \vec{x})$ as follows:

$$\psi(t, \vec{x}) = \frac{1}{\sqrt{\hbar N}} \sum_{i=1}^{N} \varphi_i(t, \vec{x})\,, \tag{30}$$

where $\varphi_i(t, \vec{x})$ denotes the function (28) for the i-th trial. To show that $\psi(t, \vec{x})$ plays the role of probability density $\rho(t, \vec{x})$, let us consider a small volume ΔV with the center $\vec{x}$ and calculate the following quantity:

$$\Delta V \rho = \int_{\Delta V} d^3x \, |\psi|^2 = \frac{1}{\hbar N} \left(\sum_{i=1}^{N} a_{ii} + S \right),$$

where

$$S = \sum_{i \neq k = 1}^{N} a_{ik} \,, \quad a_{ik} = \frac{1}{2} \int_{\Delta V} d^3x \left(\varphi_i^* \varphi_k + \varphi_k^* \varphi_i \right).$$

If one takes into account the normalization condition (29), one gets

$$\sum_{i=1}^{N} a_{ii} = \hbar \, \Delta N, \quad \Delta V \rho = \frac{1}{\hbar N} (\hbar \Delta N + S)\,, \tag{31}$$

where ΔN is the number of trials, for which the center of our soliton can be observed in the domain ΔV. Taking into account the arbitrariness of initial conditions for solitons φ_i in the i-th trial, one concludes that the quantities a_{ik} for $i \neq k$ can be considered as independent random variables with zero mean values. Therefore, in view of Chebyshev's inequality [7], one can estimate the probability for $|S|$ to surpass $\hbar \Delta N$ as follows:

$$P(|S| > \hbar \Delta N) \leq \frac{1}{(\hbar \Delta N)^2} \left\langle S^2 \right\rangle = \frac{1}{(\hbar \Delta N)^2} \sum_{i \neq k = 1}^{N} \left\langle a_{ik}^2 \right\rangle . \tag{32}$$

However, the functions φ_i and φ_k can overlap only in the domain of the volume V_0, that permits one to deduce from (29) and (32) the estimate:

$$\langle S^2 \rangle \le \alpha\, \hbar^2 \frac{\Delta N}{\Delta V} V_0 \Delta N\,,$$

where $\alpha \sim 1$ stands for the ``packing" coefficient. Thus, (32) implies the estimate

$$P(|S| > \hbar \Delta N) < \alpha \frac{V_0}{\Delta V} << 1\,, \tag{33}$$

whence with the probability close to unity the quantity $\rho = |\psi|^2$ can be interpreted as the probability density w in accordance with the Born's hypothesis.

Applying now the definition (30) to the Young's experiment, one gets

$$\psi = \sum_{j=1}^{n} \psi_j, \quad \psi_j = \frac{1}{\sqrt{\hbar N}} \sum_{i(j)} \varphi_i\,, \tag{34}$$

where indices $i(j)$ correspond to the solitons passing through the *j*-th slit. The formula (34) expresses the well-known Dirac's superposition principle that permits one to obtain the explicit structure of the "influence" function w_{jk}. In fact, inserting (34) into (27), one gets

$$\sum_{j,k=1}^{n} w_{jk} = \left| \sum_{j=1}^{n} \psi_j \right|^2 = \frac{1}{2} \sum_{j,k=1}^{n} \left(\psi_j^* \psi_k + \psi_k^* \psi_j \right) \equiv \sum_{j,k} s_{jk}\,. \tag{35}$$

Substituting $w_{jk} = s_{jk} + s_j \delta_{jk}$ into (35), one infers that $\sum_{j=1}^{n} s_j = 0$. Assuming that

$$s_j = s - \sum_{k=1}^{n} s_{jk}\,,$$

one derives the relation

$$s\, n = \sum_{j,k=1}^{n} s_{jk} = |\psi|^2$$

implying the structure of w_{ik} :

$$w_{ik} = s_{ik} + \delta_{ik} \left(\frac{1}{n} |\psi|^2 - \sum_{j=1}^{n} s_{ij} \right). \tag{36}$$

Let us now discuss the problem of measurement of some physical observable *A*, which can be represented as a bilinear form, in view of E. Noether's theorem, if the generator $\hat{M}_A$ of the corresponding symmetry group is given:

$$A = \int d^3x \pi i \hat{M}_A \phi = \int d^3x \varphi^* \hat{M}_A \varphi. \tag{37}$$

Applying (37) to the *j*-th trial, one can write the mean value over *N* trials as follows:

$$\langle A \rangle = \frac{1}{N} \sum_{j=1}^{N} A_j = \frac{1}{N} \int d^3x \sum_{j=1}^{N} \left(\varphi_j^* \hat{M}_A \varphi_j \right). \tag{38}$$

Using the definition (30) of the wave function and the estimate (33), one can derive from (38) that with the probability close to unity

$$\langle A \rangle = \int d^3x \psi^* \hat{A} \psi, \quad \hat{A} \equiv \hbar \hat{M}_A\,. \tag{39}$$

As can be seen, the formula (39) expresses the quantum mechanical rule for calculating the mean value of some observable A generated by the Hermitian operator $\hat{A} = \hbar \hat{M}_A$.

What will change if one considers the case of K particles - solitons? In this case a new definition of the wave function in the configuration space of $3K$ dimensions should be given as generalization of (30):

$$\psi(t,\vec{x}_1,...,\vec{x}_K) = (\hbar^K N)^{-1/2}\sum_{j=1}^{N}\prod_{k=1}^{K}\varphi_j^{(k)}(t,\vec{x}_k), \quad (40)$$

where $\varphi_j^{(k)}$ denotes the soliton configuration of the form (28) for the k-th particle - soliton in j-th trial. In this case the mean value of some observable A will be given by similar formula:

$$\langle A\rangle = \int d^{3K}x\psi^*\hat{A}\psi, \quad \hat{A} \equiv \sum_{k=1}^{K}\hbar\hat{M}_A^{(k)}.$$

It is worth-while now to explain, on the basis of the definition (40), the spin - statistics correlation. Suppose that all K particles are identical, that is $\varphi_j^{(k)} = \varphi_j(\vec{x}_k - \vec{r}_k)$, where $\vec{r}_k$ denotes the center of the k-th particle - soliton and functions φ_j belong to the irreducible representation $D^{(J)}$ with the weight J of the rotation group $SO(3)$. One can also identify the weight J and the spin of our particle - soliton. Let us now choose two arbitrary particles with the centers $\vec{r}_1, \vec{r}_2$:

$$\varphi_j^{(1)} = \varphi_j(\vec{x}_1 - \vec{r}_1), \quad \varphi_j^{(2)} = \varphi_j(\vec{x}_2 - \vec{r}_2),$$

and perform their transposition. First of all we rotate our two-particle system over angle π around the mediatrice of the central line $\vec{r}_1 - \vec{r}_2$. However, in view of extended character of our particles - solitons, to restore the previous state of the system it would be necessary to perform two additional rotations over angle π around particles proper axes parallel to the mediatrice. The latter operation being equivalent to the relative rotation of particles over angle 2π, it implies the multiplication of ψ by $(-1)^{2J}$. Therefore, the wave function should be symmetrical under transpositions $1 \Leftrightarrow 2$ for integer J and anti-symmetrical - for semi-integer J. As can be seen, this fact corresponds to the well-known spin - statistics correlation in quantum mechanics.

4. Conclusions

We show the effectiveness of the 16-spinor realization of the popular in particle physics Skyrme and Faddeev models for the description of baryons and leptons as topological solitons. It is worth-while to underline that the gravitational field plays an important role in our model, since the wave - particle duality principle of quantum mechanics has the gravitational origin and the vacuum excitation ξ can be considered as a tool for constructing wave functions in the special stochastic representation [5, 6] in accordance with the definitions (30) and (40). On the other hand, at small distances, due to the inclusion of the Kretschmann invariant containing higher derivatives of the metric tensor, the gravity becomes strong and provides the existence of topologically nontrivial field configurations with closed string structure [2].

References

[1] Skyrme T H R 1962 *Nucl. Phys.* **31** p 556
[2] Faddeev L D 1973 *Rep. of Ac. Sci. USSR* **210** p 807
[3] Rybakov Y P 2011 *Phys. of Atomic Nuclei* **74** p 1073
[4] Cartan E 1966 *The Theory of Spinors* (Paris: Hermann)
[5] Rybakov Y P 1974 *Found. of Phys.* **4** p 149

Journal of Physics: Conference Series **731** (2016) 012012 doi:10.1088/1742-6596/731/1/012012

[6] Rybakov Y P 2009 *Dynamics of Complex Systems* **3** p 3
[7] Loeve M 1960 *Probability Theory* (Princeton, New Jersey: van Nostrand D company inc.)

Institute of Physics
Dirac House, Temple Back
Bristol BS1 6BE UK

ISSN: 1742-6588
ISBN 978-1-5108-2729-5

CURRAN ASSOCIATES INC.
proceedings.com

Applications of Novel Scintillators for Research and Industry (ANSRI 2016)

Journal of Physics: Conference Series Volume 763

Dublin, Ireland
11 – 13 May 2016

IOP
Institute of Physics